I0493022

WEIGHING THE WORLD

WHY MASS SPECTROMETRY IS CHANGING HOW WE DO EVERYTHING

Benjamin C. Orsburn, Ph.D.

Copyright © 2014 Bench to Bedside Press

All rights reserved.

ISBN: 1499635753
ISBN-13: 978-1499635751

1 INTRODUCTION

A single fiber on a suspect's shoe links him to a crime scene. A prominent athlete fails a drug test for a new performance enhancing drug despite testimony that he never took steroids. A Rover slowly moves across Mars looking for chemical evidence that life once existed in that barren wasteland.

These are applications of a science called mass spectrometry that most of us are aware of through the news and television dramas. However, these are only a few of the hundreds of applications of this technology that are currently being explored in what is easily the fastest growing field in all of science. Many of the applications are medical, with thousands of physicians world-wide beginning to use the mass spectrometer (MS) for more rapid and sensitive diagnoses in cancer and other diseases. Just as many applications range the gamut of the sciences from detecting the nuclear enrichment capabilities of suspect nations to determining the optimal time to change the oil in a diesel locomotive. This technology is still virtually in its infancy, but already it has helped us to live in a safer and cleaner world.

This work is a small sampling of the current use of this science. I will jump from geology to vaccine research to space exploration in order to provide you with a glimpse into what we are doing and where we are going with the mass spectrometer.

I encountered my first mass spectrometer when I was 22 years old and working in the clinical chemistry department at the Johns Hopkins Hospital. Even though I was trusted to operate multi-million dollar blood analyzers, I was not on the select list of senior clinical lab technicians who had a password for the mass spectrometer, which I will refer to as mass spec, from now on in this book. Like most people out there, nothing makes me want to do something like being told that I can't do it. When I had saved up enough money to begin working on my Ph.D. full-time, I practically jumped at the opportunity to operate the two mass specs which had been donated to the Virginia Tech biology department. The majority of my doctoral work was performed on these two instruments, where I studied small molecules, compounds from bacterial spores, and even big proteins throughout the next 4 years. After finishing school, I did two postdoctoral fellowships and I obtained both jobs almost solely on the basis that I could operate a mass spec.

My plan was always to be a disease researcher. The mass spec was one tool that I planned to use to develop some new treatments for some disease or the other. During my two postdoctoral fellowships I discovered that operating a mass spec is really what I'm good at. Fortunately, there are worse things than being really good at operating the tool that is rapidly becoming the most important one nearly any science.

This text is designed to be an intro into the field of mass spectrometry, a field that I am very passionate about. I have written this in the hopes of making the field I love approachable by the average reader. To be perfectly honest, I really wanted to write something that I thought my Dad would enjoy. (My Dad is a smart guy, but he isn't a scientist.) To facilitate this end, I have attempted to leave out unnecessary nomenclature and concepts that I do not believe directly affect the facts. I hope that you will enjoy this text and feel some level of the enthusiasm that I have for the field I was so lucky to stumble into so many years ago.

2 CURIOSITY

On November 26, 2011, the Curiosity Mars rover was launched to begin its two year mission of exploring the surface of our neighboring planet. It was a big deal for dozens of reasons. It was the biggest object ever sent to Mars, it contained scientific equipment that represented the first joint Russian-U.S. project on Mars and on and on. The big mission for Curiosity, however, is to determine two main things – is there life on Mars and are the necessary resources present for humans to move there either for the short or long term? Sure, there are other issues, but these are the really interesting ones, at least for this author!

Okay, so how do you determine if there is life on another planet? Well, when Viking 2 landed in 1976, it did it this way: scoop up a bunch of

Martian soil, pour in water and a bunch of nutrients and see if it gives off gas. The result was positive. This bears repeating, I think. The first ever test to look for microbes growing in the Martian soil came back positive for life on Mars. Oh yeah, and all of the elaborate negative controls came back negative. Other tests, however, did not substantiate this finding so we threw out the results of the initial experiment as an anomaly and the Viking 2 team walked away saying that there is no life on Mars. The results of the Viking 2 experiments have been controversial to this day. More on this later.

The scientists and engineers who built Curiosity went another direction with their design.

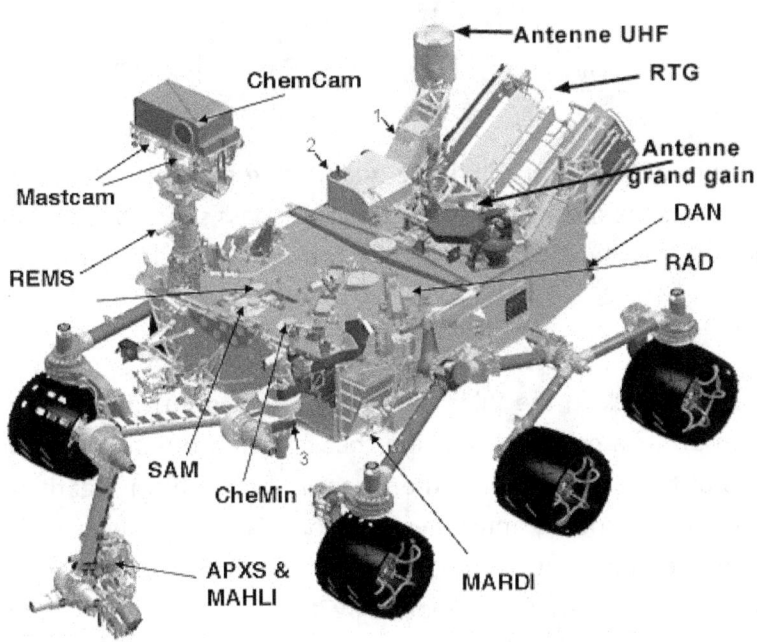

Curiosity is a science first mission. As such, it carries over 10 times the mass of total analytical instruments compared to any of its predecessors. The center piece is called SAM, which is short for Sample Analysis at Mars. The following illustration is a close up schematic of SAM.

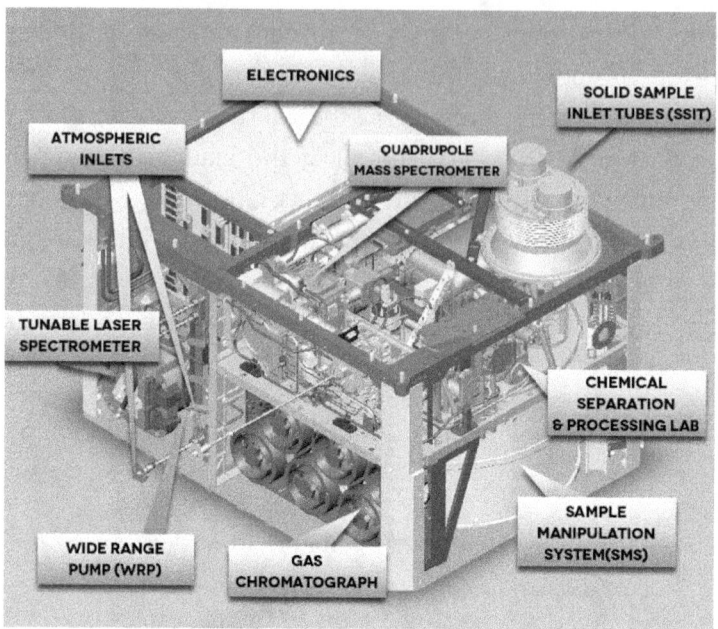

And at the core of SAM you will find a quadrupole mass spectrometer. These are neither the most sensitive or accurate mass specs, but as they are composed of 4 beams of steel, they are known to be extremely rugged.

Let's take a step back to the Viking 2 results. That positive result has been written off as a chemical reaction that occurred due to inorganic components of the Martian soil, rather than as an organism breathing. Of the many jobs Curiosity must perform, one is to test the soil for inorganic compounds that could have caused that positive result. Identification of one of these compounds with the mass spectrometer will confirm that the original test for life simply didn't consider enough variables. If Curiosity cannot identify inorganic compounds that could produce rapid gas formation, then the 40 year old results from Viking 2's first experiment may still stand.

3 FORENSICS

Thanks to television dramas such as CSI, NCIS, and Dexter, forensics has become a common word and concept in today's world. We are now very familiar with the role that labs play in assisting law enforcement investigations. Mass specs play a key role in these facilities and have been one of the earliest uses of this technology outside of the physics lab. The value of the mass spec is particularly obvious in NCIS, as it is the primary tool of Specialist Abby Sciuto.

Now, the science in television dramas such as these is under constant scrutiny by the scientific community. Facts are often omitted or exaggerated for entertainment value, but there is no question that the mass spec plays a key and central role in forensics labs.

On November 10, 2012 a home exploded in a suburb of Indianapolis, Indiana. The explosion destroyed neighboring houses and killed a couple next door who had come home for work early. The owners of the exploding home were quickly arrested and charged with over 50 counts of arson. The homeowners have contested these charges and testify that they had no hand in the explosion, and that it was the result of an unknown natural gas leak.

One of the first uses of the mass spec by law enforcement was in the confirmation of suspected arson cases. Between 1987 and 1997, there were over 500,000 confirmed cases of arson in the United States alone. This accounted in over $3 billion dollars in property loss. In most cases, arsonists use chemicals known as accelerants in order to get the fire burning as rapidly as possible. Traditionally, arson was confirmed the old-fashioned way, by the trained noses of fire fighters and marshals.

Unfortunately, prolonged exposure to burning debris and smoke has a way of lowering one's sensitivity to smells. After exposure to hundreds or even thousands of fires throughout a veteran's career can sometimes impair one's ability to smell gasoline remnants a little limited. One also has to wonder if the efficacy of diagnosing arson falls during cold and flu season.

In the 1980s work began at several forensics labs to develop solid methods for the detection of accelerants with mass spec. In 1995 researchers from the Institut de Police Scientifique and the Forensic Science Unit of Glasgow compiled data from these analyses into a database available to police world-wide that contained information for the complete identification of 40 common accelerants. An expanded version of this database is still currently in use and is used during investigations of nearly every home or commercial fire. A key piece of evidence in the trial in Indianapolis will, undoubtedly be, the results of the mass spec analysis.

Another use of the mass spec in forensics is in the verification of evidence linking a suspect to a crime. One major piece of evidence is single fibers from cloth and upholstery. While fibers can be manufactured in a wide variety of tolerances and are often not directly admissible in court, factories that make fibers have a wide variety of dyes to choose from. Researchers at Oak Ridge National lab have demonstrated that the profile of these dyes in these fibers are so unique that one fiber from one piece of material has a unique fingerprint that can link it directly back to the factory – or to the crime scene. In this procedure, the suspect is searched for fibers and these are manually checked for color and texture by investigators for similarity to those at the crime scene. Fibers that appear similar are separated and the dyes are removed by treating them with solvents that will extract the color. Mass spec analysis can verify that the fibers contain exactly the same kinds and amounts of dye, thus linking the suspect to the crime scene.

While we're on the topic of fingerprinting, we all know that our fingerprints are unique. What if they don't just look unique? Prosolia is a company that makes a living on the fact that fingerprints aren't just physically unique, they are also chemically unique. A Prosolia device can look at an individual fingerprint, match it to the suspect, and tell whether that person has touched drugs or explosives. The sensitivity of this assay is to the point that tests have shown the successful identification of explosive residues in the fingerprints of test subjects after considerable time periods. The instruments are also relatively inexpensive and are showing up in police labs around the world.

These are just a few of the many applications of mass spectrometry in forensics. As one of the first areas outside of pure scientific research to adopt these instruments and techniques, forensics has been front-and-center to enjoy the advances in instrumentation, processing software, and sample preparation techniques as this science has developed.

References:

1) Clodfelter, R.W. Comparison of decomposition products from selected burned materials with common arson accelerants. Journal of Forensic Science 22, no. 1 (1977): 116-118

2) 'The Richmond Hill Explosion' *Wikipedia. The Free Encyclopedia* http://en.wikipedia.org/wiki/Richmond_Hill_explosion

3) Desorption Electrospray Ionization of Explosives on Surfaces: Sensitivity and Selectivity Enhancement by Reactive Desorption Electrospray Ionization Ismael Cotte-Rodríguez,Zoltán Takáts,Nari Talaty,Huanwen Chen, and, and R. Graham Cooks Analytical Chemistry 2005 77 (21), 6755-6764

4 SOME HISTORY OF MASS SPECTROMETRY

It is often pretty hard to fully attribute a scientific development to a single person. Science often works best when one person makes slight improvements on another scientist's studies. It is the same with the invention of mass spectrometry. A long line of chemists studying the effects of electricity and magnets on atoms of different kinds led Sir Joseph Jon Thomson to the invention of the first mass spectrometer. This was just one of his many accomplishments in science, including the discovery of the electron and proving that it was indeed a particle.

Thomson's mass spec looked something like this:

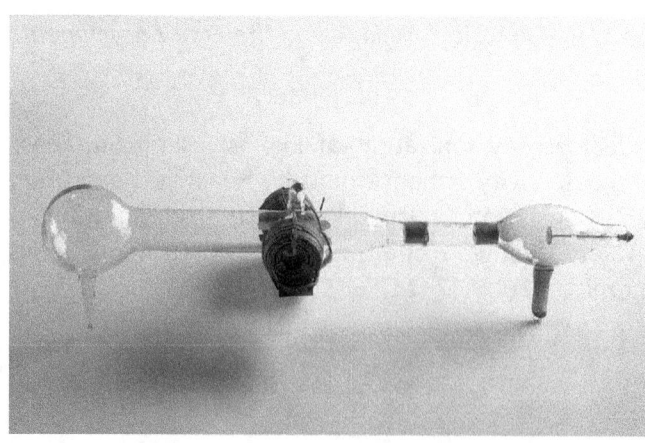

(Image courtesy of the Science Museum of London/John Cummings)

What he was able to discover with this contraption of glass and electromagnets is that by shifting the polarity of gas moving in a vacuum he could deflect its direction.

For example: in a similar device to this one, gas molecules could be directed to beam straight through the device as so:

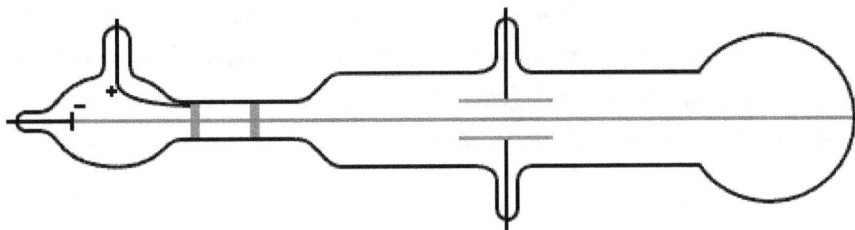

However, by adding an electric charge at the center places, Thomson could force the gas ions to move in other directions depending on where the charge was placed on the plates, as shown below:

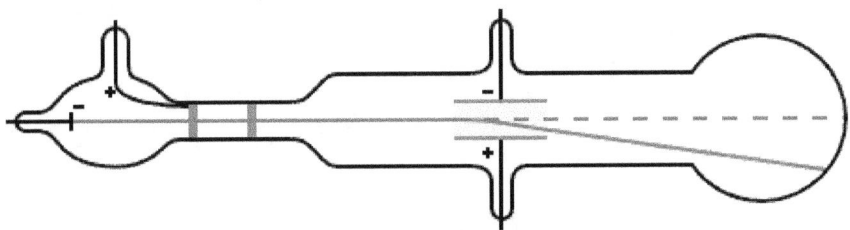

Using devices similar to these, Thomson was able to separate different isotopes of Neon gas. He discovered that using the same voltage would separate neon in two different directions and that the heavier isotopes would shift more.

Thomson's student, Francis Aston, took this discovery a step further in his research by using devices similar to these on different types of gases. It was soon apparent that this characteristic was not unique to Neon gas. He discovered that elements of different masses behaved differently when affected by the electromagnetic field and that this difference was always directly proportional to the mass of the element. With this device he was able to discover isotopes of several elements. For years the basic design of the mass spectrometer remained unchanged. Chemists at this time could reuse similar devices to study different elements and build our basic understanding of chemistry. In fact, instruments similar to these are still in use today for specific applications of physics and basic atomic chemistry.

For their work in this field, both Thomson and Aston received separate Nobel Prizes. Thomson received the Nobel in Physics in 1906. Aston received the Nobel in Chemistry in 1922. At the point of this writing 5 Nobel Prizes in Chemistry have been awarded for the development or application of new mass spectrometers.

5 MODERN MASS SPECTROMETRY

In the last chapter I talked about the bending of ions being different depending on their mass. In 1946, William Stephens reported the idea of a time-of-flight (TOF) mass spectrometer.

The image below is a crude illustration of TOF. It is, in my opinion, both the easiest mass spectrometer to understand and the easiest to draw in PowerPoint.

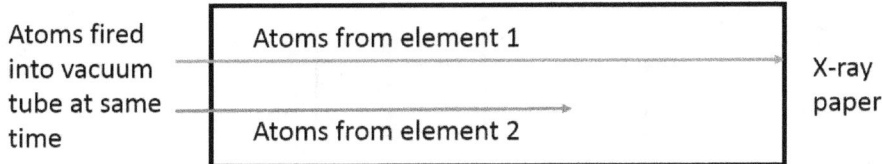

In this scenario it makes some sense, right? We're shooting two objects into a vacuum tube. We have a sensitive way of telling when the atoms reach the end of the tube. The larger or heavier atoms takes longer to get there. The bigger one accelerates more slowly and then ends up getting there after the smaller one.

An important point to keep in mind is that it's pretty tough to detect a single atom. But what you can do is fire a whole lot of each atom into the vacuum tube at the same time. For example, say we fired 10 atoms from element 1 and 5 atoms from element 2 into the tube.

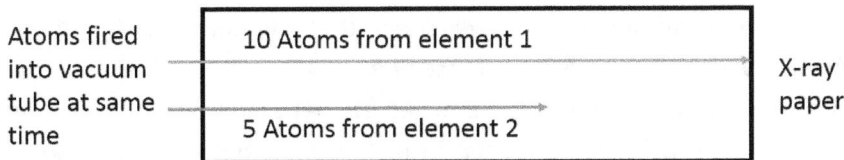

Now, if we were just hitting using the atoms to make a mark on a piece of paper, it would be really hard to tell if 1 or 1,000 atoms hit in that point. What we need here is an electronic detector that can measure can tell by a shift in the electronic current about how many atoms hit it at once. Fortunately, a kinda famous guy named Michael Faraday had created something like that almost a century earlier.

If we take the example above and we have a way of measuring the number of atoms to strike the detector at the end, we can plot this data and we end up with something like this:

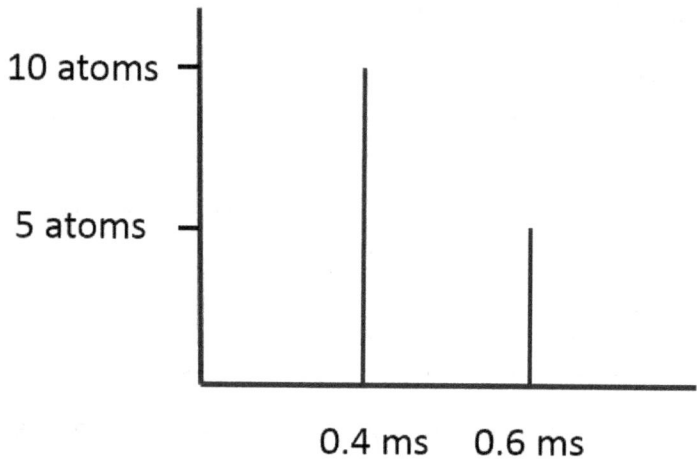

On the X-axis, we have the amount of time that it takes for the atoms to hit the detector. On the y-axis, we have the number of atoms. In my example, I'm saying that the 10 atoms can make it to the detector and be recorded in 0.4 milliseconds. The atoms from the other element all get there 50% slower and end up taking 0.6 milliseconds (ms).

The units in use here, number of atoms per unit time is kind of useful, but it turns out a more valuable piece of information is the mass of each atom. For example what if we knew that element number 1 was oxygen. Oxygen has an atomic mass of around 16 (15.999) atomic mass units (amu). If we didn't know what element number two was, we might be able to figure it out based on the fact that oxygen could get there in 0.4 ms and our unknown compound took 50% longer. Simple arithmetic tells us that 50% more than 16 is 24. So it's possible that the unknown element is Magnesium, with an atomic mass around 24.

If we re-plot our graph with this in mind, we can come up with something like this.

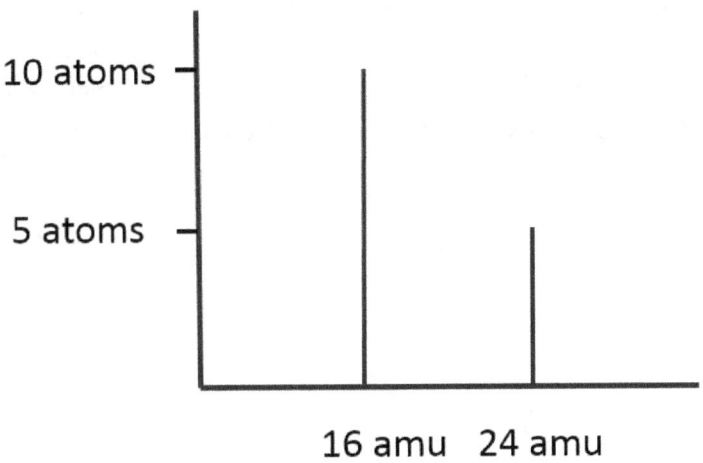

What we've generated here is this book's first mass spectrum.

Throughout history, the developments that have led us to modern mass spectrometry instruments have generated numerous Nobel Prizes in Chemistry and Physics. Think whatever you want of the Nobel committee, but they don't give out the Science and Medicine ones for small accomplishments. What I'm getting at is that this is an extremely simplistic view of the mass spectrometer, but this is also how they work on an extremely basic level. To this day, virtually every one of the thousands of mass spectrometers on earth run a series of standards in order to calibrate the instrument. They typically range from low mass to high mass, so we know that what masses correlate to what features of the way the atoms move under vacuum. Some modern mass spectrometers, such as the Orbitrap instruments I'll discuss in later chapters can maintain their calibrations for weeks or even months.

Other instruments must be calibrated much more often, with some requiring a known standard is put in with every unknown sample.

While measuring the mass of elements and the relative numbers of each one around is a useful thing to do, the real power of the mass spectrometer comes from the fact that much larger objects can be measured. Recent advances have allowed the accurate mass spectrometry measurements of intact proteins from human beings and even the measurement of protein complexes with masses as high as one million amu. All of these advances will be discussed in later chapters.

6 DRUGS IN SEWAGE

People all over the world take drugs. People all over the world even take drugs that their government tells them they aren't allowed to take. Government bodies and health organizations are very interested in these facts and being able to quantify how many people are taking what drugs, and when, can be an extremely valuable bit of information.

Interestingly, many of the people in developed countries who take drugs also expel waste into depository devices that are connected to centralized municipal facilities; that is, they commonly urinate into bathrooms connected to sewers. Enterprising scientists all over the world have used these relationships to determine exactly who is doing what drugs and when.

This is how it works:

Our enterprising (and extremely committed) scientist takes sewage samples at locations and times of interest and takes it to her lab.

The lab then filters out the extremely gross bits. They probably also centrifuge it at to generate forces thousands of times stronger than earth's gravity to filter out the moderately gross bits. The remaining liquid may also be treated with chemicals that will kill the things living in the solution that are known to not interact with the drugs being analyzed.

Now, even after this degree of filtration and analysis there is still lots of ~~shit~~ chemical compounds in the clarified sewage. The mass spectrometry portion of this analysis must be extremely precise.

In order to do this analysis, the lab must start by analyzing a pure sample of the drug they are studying. In the case of cocaine, the compound that is studied is called cocaethylene. It is formed when cocaine and alcohol are combined in the body, a practice that is believed to be common. The lab first optimizes their instrument to obtain the best possible sensitivity on pure cocaethylene.

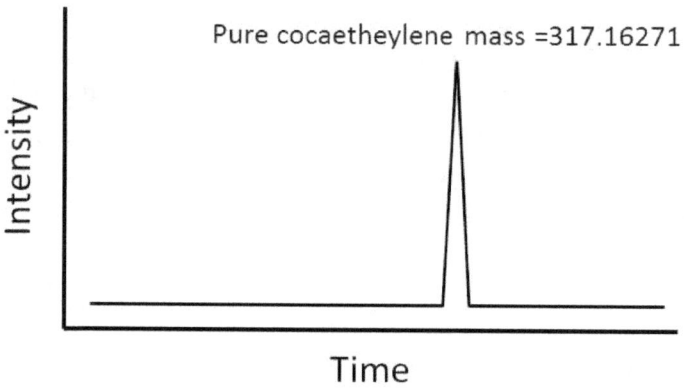

In most experiments, the compound is first ran out on a column using HPLC which is short for High Performance Liquid Chromatography. (We long ago gave up on Low Performance Liquid Chromatography). In this technique liquid containing different analytes passes through a porous column. Different compounds interact with the material in the column to different degrees. Some compounds come off the column quickly, while others take longer. This is a very common procedure in nearly every analytical lab.

In the case of our example, we'll say that our pure cocaethylene comes off our column at 7.5 minutes. By connecting our HPLC column to our mass spectrometer, we can get the exact mass of every compound as it comes off the column. In the case of cocaethylene, we are looking for exactly 317.16271. The coupling of an HPLC system with a mass spec is often referred to as LC-MS for short.

Once the system has been fully optimized on the known standard, the brave researchers take a deep breath and inject their cleaned up sewage sample onto their expensive, extremely well maintained, and sensitive LC-MS system.

The results may look something like this

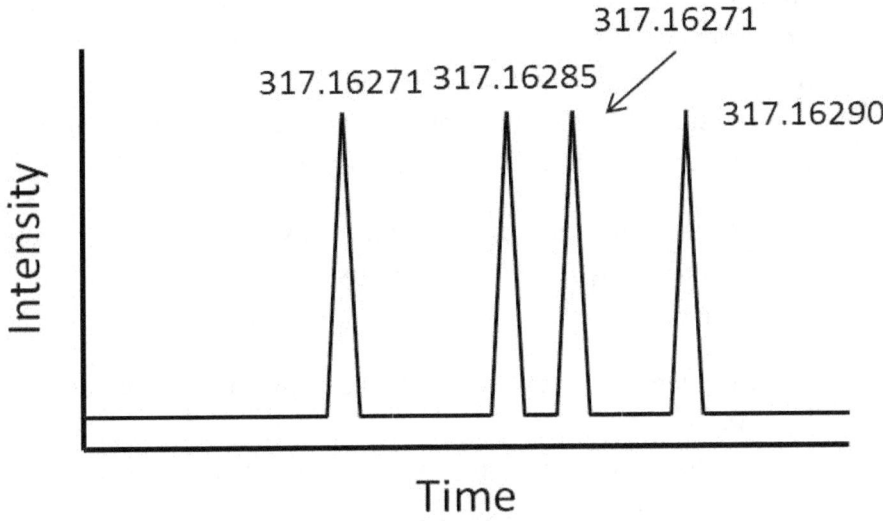

This is why the HPLC system is used in conjunction with the mass spec. We need multiple points of evidence to accurately say that the compound we are quantifying is really cocaethylene. The first peak is a something that has the exact mass of our drug compound, but it comes off the HPLC at the wrong time. The second peak comes off very close in time to where we would expect to see our compound of interest, but the mass is slightly off. Only the third peak comes off at the correct time and has the correct mass.

Now, it is worth stating at this point that the difference between the second and third compounds is extremely small. It is only 0.00014 Da, or 1/10000 of the mass of a proton. Only extremely high resolution mass spectrometers are capable of determining mass differences this small. Only the Orbitrap system mentioned in another chapter is really capable of measurements this precise. Nearly every other mass spectrometer on earth would give identical values for all of these peaks, probably something in the range of 317.16. On these instruments, a third degree of certainty must be used. These instrument use techniques called single or multiple reaction monitoring (SRM or MRM) that essentially require two mass spectrometers working together. The first filters for ions within a narrow mass range and the second studies the fragments of those ions in order to match the fragments to those from the pure compound of interest. Essentially, however, the techniques end with the same results. Accurate measurement of the amount of the drug of interest in the sewage sample.

What have we discovered from all of this sewage studying? There are a number of interesting observations, including but certainly not limited to:

- The Dutch do not appear to use any more drugs than the international average
- Amphetamine usage at universities all over the U.S. and U.K. spike during finals week
- Having the Super Bowl in your city does not increase the amount of drugs consumed by that city. The amount of cheap, terrible, American light beer, however, is off the charts (I made that part up)
- Cocaine and ecstasy usage is much higher in cities and that usage is much higher on the weekends (though this first statement may have something to do with the fact us country folk sometime do our business outside...)

Now, many of these observations seem obvious. But these aren't things that have, or really could be, quantified in any measurable way without some daring researchers and without mass spectrometry. By the way, I made these studies seem like an anomaly. They aren't. A Google Scholar search of the terms "drugs in sewage" will lead you to hundreds of articles detailing everything from party drugs to antibiotics to chemotherapeutics and their identification and quantification in sewage all over the world.

References:

1) Investigation of drugs of abuse and relevant metabolites in Dutch sewage water by liquid chromatography coupled to high resolution mass spectrometry. Bijlsma L , Emke E, Hernández F, de Voogt P. Chemosphere. 2012 Nov;89(11):1399-406

7 JOHN FENN

In the previous chapter I described how liquid chromatography (LC) systems coupled to mass specs can lead to massive increases in the amount of data that we can look at per experiment. We couldn't always connect these two devices. The technology that allowed us to do this is called electrospray ionization and resulted in a Nobel Prize for Professor John Fenn in (2002).

Fenn's story is almost mythical among academic researchers. To put it bluntly, Fenn was kind of a badass. What we need right now, however, is a bit of background into the story.

Becoming a tenured professor in the sciences is an incredibly difficult thing to do. First you need a Ph.D., then you need to perform postdoctoral research and publish some high level research to be even considered for an Assistant Professorship. Once you've obtained this position, the hard part has often just begun. You need to prove to your University that you can successfully balance the classes you teach with managing a full research program of undergraduate volunteers, graduate students and post docs that is fully self-sufficient due to all the grant money you bring into the school. Then, and only then, do you have a shot at making it to being a tenured professor.

All this work pays off at this point. Tenure gives you the option of life long employment with your university. There is still enormous pressure to teach and publish and bring in grant money, but you know that even if you have a few bad years where you only contribute a little to the scientific fields, or only bring in a few hundred thousand dollars in grant money, then you won't be thrown out on the street.

From a financial perspective, most Universities prefer professors who are either killing themselves for tenure, or who have made it within the recent past. This is due to the amount of grant money professors can bring in when their back is against the wall. Tenured professors who publish less and bring in less money often become less attractive from a purely financial perspective.

Dr. Fenn was one such case. He wasn't racking in grant money and he wasn't publishing all that much. The University he worked for began cutting his resources in order to make room for younger, more productive professors. Eventually Fenn was down to a tiny lab space and no graduate students. The University was hoping that he would soon retire and move on.

During this time and working almost entirely on his own, Fenn made the discovery that would lead to his eventual Nobel Prize in Chemistry and world-wide recognition. Fenn discovered that liquid flowing out of the tip of a needle at an extremely slow flow rate could be an input for a mass spectrometer if that needle was charged with about 2000 volts of electricity. The process is called electrospray ionization and is now used by virtually every mass spec on earth.

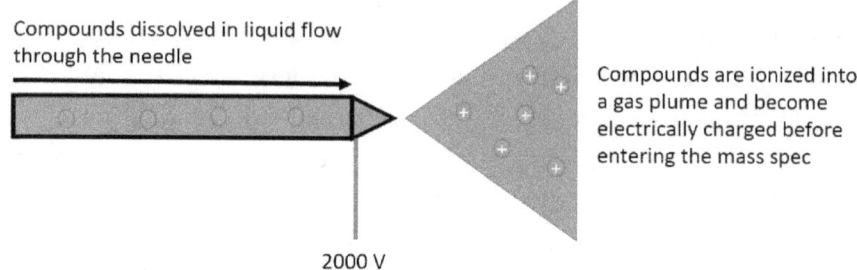

Compounds dissolved in liquid flow through the needle

Compounds are ionized into a gas plume and become electrically charged before entering the mass spec

2000 V

Now, it is fair to guess that Fenn had a good idea of the commercial value of his discovery. Most Nobel Prize winners are pretty smart. And it might be safe to assume that Fenn felt a little slighted by his University and the way they had treated him after his decades of faithful service.

It is also possible that he did underestimate the value of his discovery. In either case, when John Fenn left his University and joined Virginia Commonwealth University, he estimated the total value of the patents he had developed while at his old work as not a lot of money. Which was fine, at first.

Once the full implications of Fenn's discovery became understood, a long legal battle began with Dr. Fenn on one side and University lawyers on the other, as they worked for years to sue Fenn for their share of the value of his patents and awards. Fenn fought his old employers until he died peacefully, and of old age, in 2010. And his University finally got the money they felt they deserved.

I don't know how many labs I've been in where pictures of John Fenn have hung on the wall somewhere. Sometimes it is out of respect for the most recent Nobel Prize awarded for mass spectrometry. Sometimes it's up purely out of respect for a man who stuck to his guns and his beliefs and was pretty badass all the way to the end.

8 THE ORBITRAP

In Chapter 5 I talked about the simplest mass spectrometer, where we shoot an atom or ion from one place to another. As I mentioned, this is called a Time of Flight mass spec, or TOF. The way to improve the results from the TOF is to lengthen the flight path. For example, I'll bring back the TOF illustration from Chapter 5. What would happen if we sent the same ions into a longer flight path?

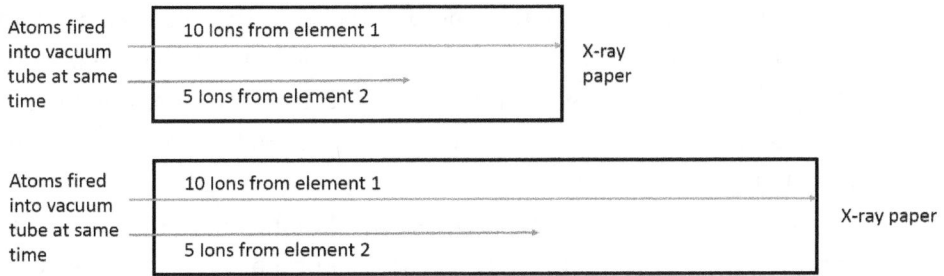

By increasing the flight path by 50%, we give the ions that much longer to separate in time and space. All of a sudden, ions that are similar in mass are separated by a larger distance. For example,

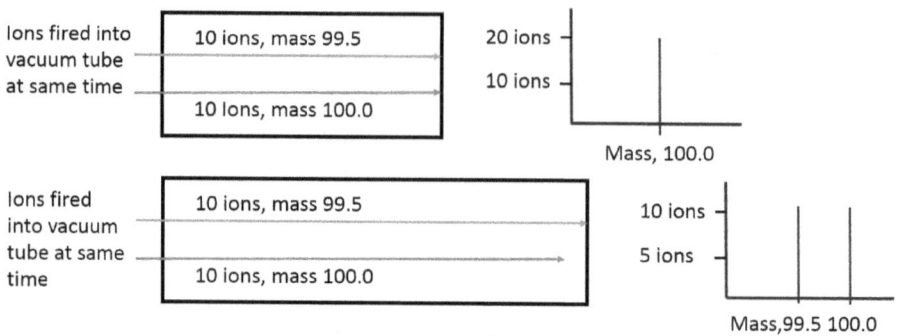

In this example, 20 ions are fired into the mass spec. 10 of them have a mass of 99.5, the other 10 have a mass of 100.0. On the short flight path (top) the flight tube isn't long enough to separate two ions that are so similar in mass. The resulting spectrum looks like there are 20 ions with a mass around 100 amu.

In the lower example, the ions have a 50% longer flight tube. This means that the ions with a mass of 99.5 get to the end of the flight tube a measureable amount of time before the ions that have a mass of 100.0. The mass spectrum is essentially "zoomed in" when compared to the spectrum from the shorter tube because this instrument can separate ions of similar mass. We can now tell that there were 10 ions of each mass. The ability of the instrument to tell the difference between similar ion masses is called resolution. It is similar to the resolution of a TV, because higher resolution on a television gives us a greater ability to see the difference between two objects by putting more pixels in-between. Here we have placed more units of mass in-between.

It would be amazing if we could just build longer and longer flight tubes. If we wanted to tell the difference between two ions that differ by 0.0001 Da, we would just build a tube that ran from one side of a big university campus to another. Unfortunately, there are strict limits to how long of a tube we can make. This is mostly due to the fact that we can't make a perfect vacuum yet here on Earth. By using multiple high output vacuum pumps we can make a great vacuum, but we can't make a space that is completely devoid of any gas. The larger the space is, the harder it is to get a great vacuum.

Take this example:

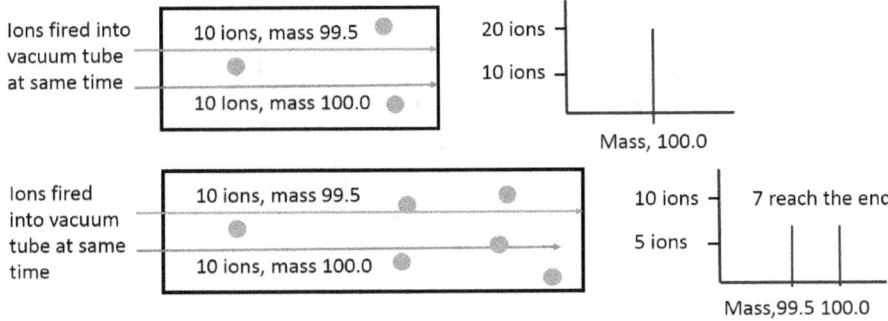

In this illustration, I've inserted circles to represent gas molecules. Every millisecond the ions are in the tube increases the chance that the ions we are firing to the end will hit a stray gas molecule. This will slow down or even stop the ion in its path. In the example of our longer flight path, we've increased our resolution, but we've decreased our signal. If we make the tube long enough to get incredible resolution we won't end up with any ions left to measure.

People attempted to combat this fundamental feature of TOF analyzers. In fact, they still do. We can accelerate the ions harder, or even accelerate them in the middle of their flight, but at the end, this is just a fundamental feature that we can't do anything about. In order to achieve the dream of both high sensitivity and high resolution, a new solution was needed.

Enter Dr. Alexander Makarov, a Russian borne physicist with a unique new idea. Dr. Makarov realized that it didn't matter whether the ion was moving in a straight line or not. The ions would still separate based on their mass. What if ions were introduced into a small chamber, and an electrical field forced the ions to move in one direction while circling a central rod.

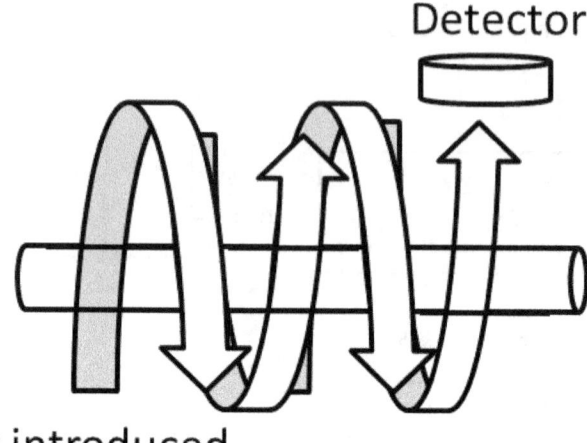

Detector

Ions introduced

In this case the ions are forced by a circular electrical pull to move clockwise around this pole. At the same time an additional force is pulling the ions toward the detector from right to left. By making the ions move in a circular path, we can dramatically increase the distance the ions have to go. As in the TOF example, we can increase the resolution by increasing the length of the rod.

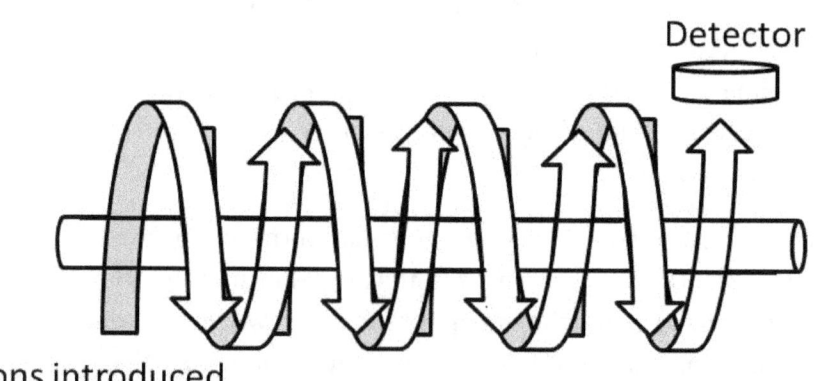

Detector

Ions introduced

Ultimately, however, we're going to run into the same disadvantage as in the TOF. It is harder to get a good vacuum in a larger space. This is where a good idea becomes a great idea. Rather than have the ions strike a detector in Dr. Makarov's new mass spec, the force that is pulling the ions to the right reverses before the ions have a chance to reach the end of the pole. All the ions then head back to the left, and back to the right, and back to the left. The ions never hit anything. They simply enter the instrument, go to the right of the vacuum and left and right again. By alternating this electric field, but by keeping the clockwise revolution of the ions intact, the flight path become long – very long. Where the longest flight path of a functional TOF ever made is around 12 feet, the flight path of Dr. Makarov's instrument, the Orbitrap, can be thousands of feet in length.

The fact that the ions travel the same path repeatedly means that the Orbitrap can also be small. Very small. Due to this tiny size, the vacuum in the Orbitrap can approach phenomenally low levels and allows the sensitivity to be extremely high compared to TOF instruments. How high? So high that many people believe that the detection of a single molecule may be possible in the Orbitrap. Due to quantum effects, however, this will be pretty hard to prove.

To illustrate the size of this powerful device, the following picture is of two different Orbitraps by my Espresso cup.

As I mentioned above, however, the ions never strike anything like an X-ray film or an electrical detector. Instead, the frequency at which the ions pass from side to side is used to determine the mass of the ions.

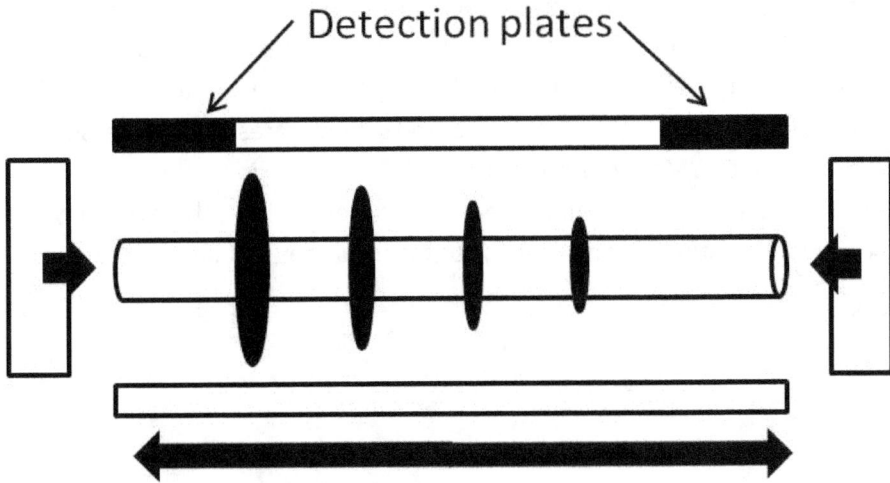

As the ions pass by the detection plates on the outside of the Orbitrap, the electrical force of their passing can be determined and recorded. The translation of the frequency of their passing can changed into the mass of the ion by way of a complex equation called the Fast Fourier Transformation. Now, this equation has a pretty grand title. It should, because the paper that it is described in is the most cited original research paper in history.

Now, I have never met Dr. Makarov. I have, however, accidentally jumped in line in front of him at a coffee maker when I didn't recognize him and I was desperately jetlagged. I do feel like I know him a little, however, thanks to a description of the history of the Orbitrap that he wrote for the magazine The Analytical Scientist in 2013. The article "Orbitrap Against All Odds" describes his struggle to not only develop the instrument, but also to convince people that it was a good idea. I have been told by people in the field that every major mass spec manufacturer passed on this new technology. After many years, he finally found a company to help develop this idea to fruition. When the first Orbitrap was released in 2004 it blossomed to near instant success and has dominated both the field of mass spectrometry and the scientific literature ever since.

One of my colleagues made the following graph showing the number of publications in the world's leading science journals that involved Orbitrap mass spectrometers and compared that to all of the TOF instruments made by all manufacturers (who presumably passed on developing Dr. Makarov's instrument). The X-axis is the number of publications per year. As someone who has a side job reviewing the scientific literature, I can assure you this trend has not changed.

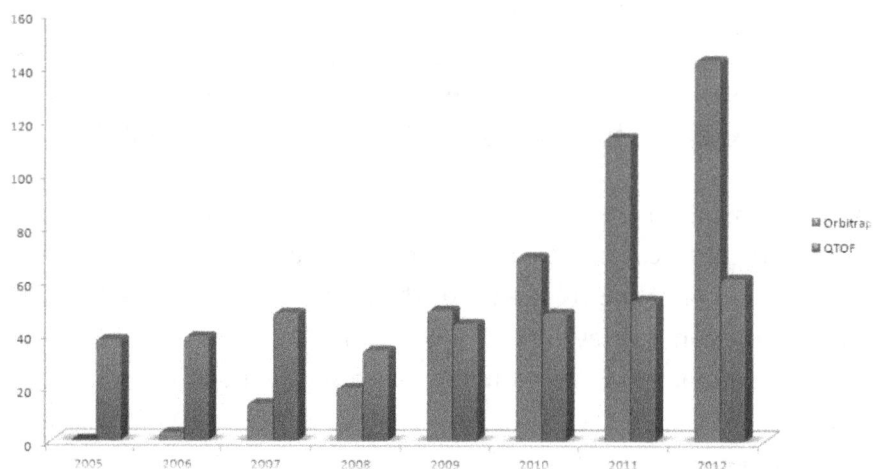

(Image Courtesy of Dr. Julian Saba)

As I mentioned in Chapter 4, 5 Nobel Prizes have been awarded for the invention or application of new mass spectrometers. The Orbitrap is the first completely new addition to the field since the ion trap was invented in over 30 years ago. It is only a matter of time before the Nobel committee acknowledges Dr. Makarov for his contributions.

Reference:
Orbitrap Against All Odds. Alexander Makarov. 2013. The Analytical Scientist, Issue 1013

9 THE HUMAN PROTEOMICS PROJECT

On June 26, 2000, Bill Clinton and Tony Blair held a joint press conference. After a decade of work led by some of the world's top scientists, the first rough draft of the Human Genome Project (HGP) was completed. This was a road map of the DNA of human beings and detailed nearly 30,000 genes that direct why one human being differs from another. It was a grand day for the medical and biological sciences and the first big success for the new science of biotechnology. We now know pretty much the exact location in our DNA where we store the directions for eye color and heritable diseases. This also opened up the discussion of genetic screens that could predict whether we would get cancer or even our own personal lifespans.

Nearly 15 years later and we've yet to see many of these predictions come to light. While the HGP was a tremendous success and the information is invaluable to some select fields, we know now that we were underestimated biological complexity. Some might argue that we were also focusing on the wrong biological molecule.

To back up a bit, DNA is where we store our genetic information. We have two strands twisted together in the famous double helix shape. One of the strands we obtained from our biological mother and the other from our father. When we have children of our own, we will pass half of our DNA down to each one of them as well. While exceptionally important, DNA has only one real job, and that is to make RNA. RNA is a simple and rapidly degrading cousin of DNA. Each gene contains the code for making one specific strand of RNA. Before that strand of RNA degrades it carries out it's one and only job – it codes for the making of a protein. This relationship between DNA, RNA and protein is so key to the biological sciences that it is called the Central Dogma.

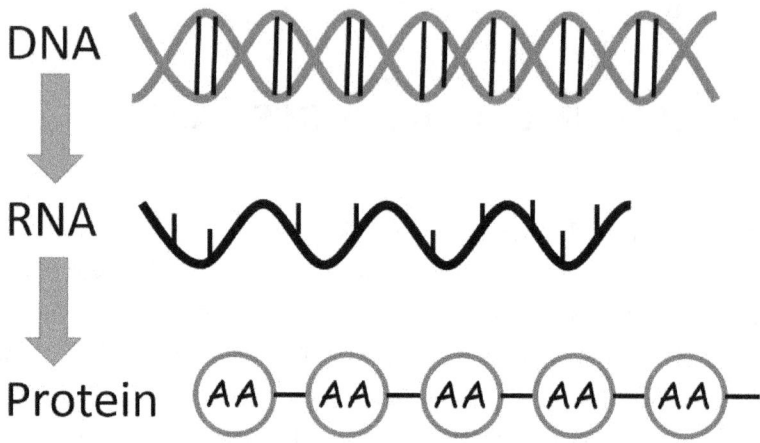

Proteins are the molecules in the cell that perform actions. Some proteins physically move things, working as little motors that bind to nutrients and push them to where they belong. Some proteins are structural, making up the fibers that keep our cells in their correct shape and make up our hair. A surprising number of proteins have the interesting job of modifying other proteins. They can do this by chopping off part of the protein or by connecting two or more proteins together, or by adding other molecules such as sugars or phosphates to the protein. Protein modification is so rampant that current estimates place the total number of possible individual protein combinations at more than 1,000,000 and possibly as high as 5 billion in human beings alone. This wouldn't be such a big deal except that when you chop off part of a protein, or bind a sugar to it you often completely change what that protein does.

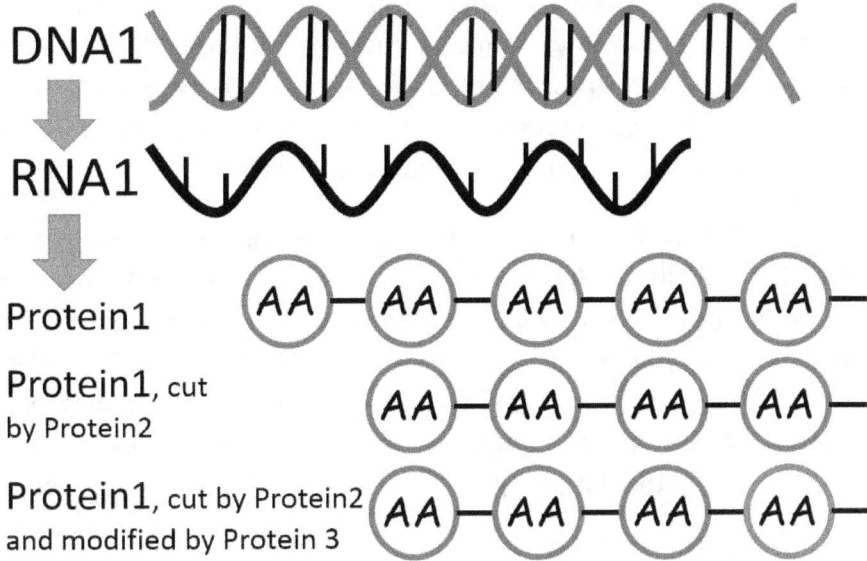

In early 2014, I saw an image that was making the rounds on Twitter among Proteomics scientist. This image really captures the difference between genomics and proteomics. The picture is shown below.

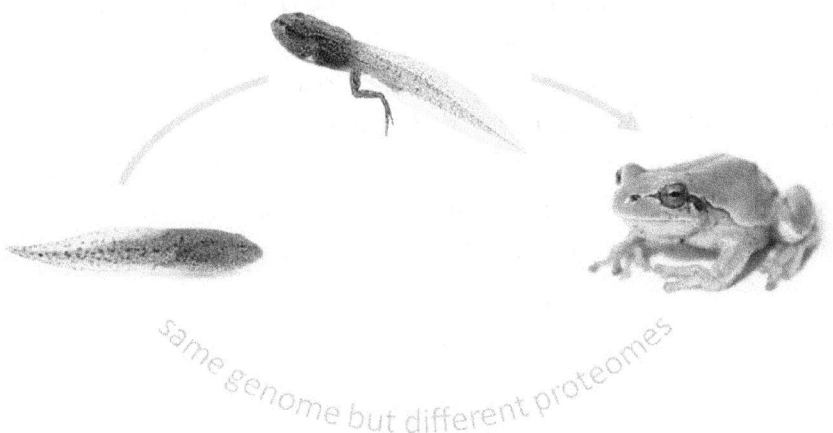

(Image courtesy of Pastel Biosciences)

At any point in the frog life cycle, the DNA is unchanged. The same base pairs are there and in exactly the same pattern, but the proteins expressed by the frog are going to be very different. As I'm not an expert on frog anatomy, I can only guess where these changes are. A good example though would be that the tadpole is expressing tadpole tale proteins. These proteins are not expressed in the adult frog. But the genes to encode both of these things are present.

This was the missing info in the excitement leading up to the release of the HGP. While we know where each gene is one every one of our chromosomes, this is only a small snapshot of the information necessary to figure out what even a single cell is doing, or about to do. In order to really understand what is happening in our cells on a chemical level, we need to have a map of all of the proteins that a human cell can make. As a play on the human genome, we refer to this as the human proteome.

In 2010, the human proteome project (HPP) was announced. Due to the size of this undertaking, the 26 human chromosomes were divided between the top research facilities in more than 17 different countries. The project is called the chromosome centric human proteome project, or CHPP. While first world nations such as the United States, Japan and France are big contributors, researchers from other nations are bringing just as much to the table. In a show of science surpassing political boundaries, research labs in Iran have volunteered to completely map the protein of the Y chromosome. Every year the Human Proteomics Organization, or HUPO, meets to hold an annual congress in one of the participating nations to update everyone on their current progress.

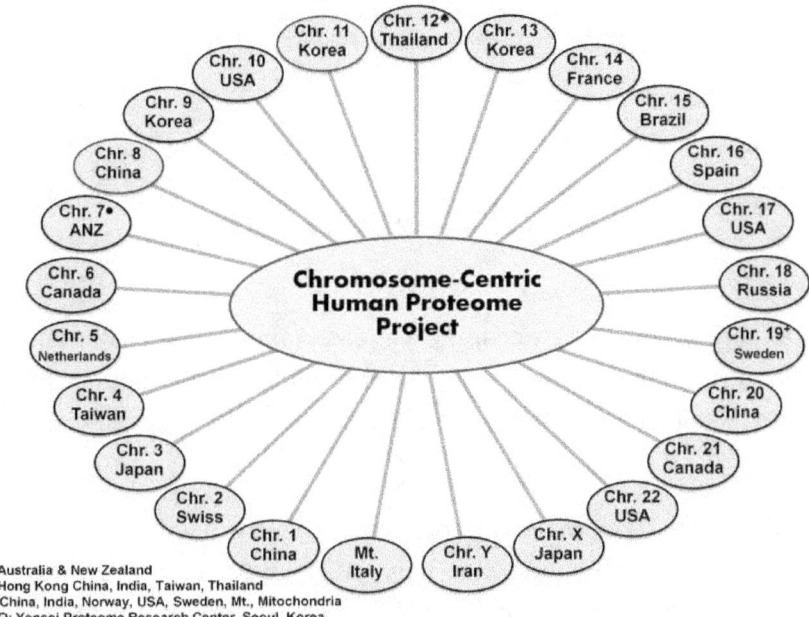

• Australia & New Zealand
♦ Hong Kong China, India, Taiwan, Thailand
+ China, India, Norway, USA, Sweden, Mt., Mitochondria
HQ: Yonsei Proteome Research Center, Seoul, Korea

During the review of this book, the first 2 drafts of the human proteome project were simultaneously announced in the journal Nature. These releases shared the cover and have been hailed as a major step forward for both the biological and medical sciences.

References:

1.) Paik, Young-Ki, et al. "The Chromosome-Centric Human Proteome Project for cataloging proteins encoded in the genome." *Nature biotechnology* 30.3 (2012): 221-223.
2.) Kim, Min-Sik, et al. "A draft map of the human proteome." *Nature* 509.7502 (2014): 575-581.

10 SHOTGUN PROTEOMICS

Wait a minute. In the last chapter we talked about how proteomics or protein mass spectrometry can reveal more information about what is going on in cells than what we can get from genomics approaches. How does this work?

Well, there are two main approaches. The so-called "shotgun" or "bottom-up" approach and the new intact protein or "top down" approach. Before we go into the differences, it is important to go into some boring details about what proteins are.

Virtually all proteins on earth are completely composed of just 26 separate compounds, called amino acids, shown in the chart below. The amino acids vary quite a bit in size, structure, and mass but they have a common structure with a carbon atom sticking off one side and a nitrogen atom sticking off the other. For the sake of brevity(?) we refer to these as the C-terminus an N-terminus respectively.

1-letter code	3-letter code	Chemical formula	Mass
A	Ala	C_3H_5ON	71.03711
R	Arg	$C_6H_{12}ON_4$	156.10111
N	Asn	$C_4H_6O_2N_2$	114.04293
D	Asp	$C_4H_5O_3N$	115.02694
C	Cys	C_3H_5ONS	103.00919
E	Glu	$C_5H_7O_3N$	129.04259
Q	Gln	$C_5H_8O_2N_2$	128.05858
G	Gly	C_2H_3ON	57.02146
H	His	$C_6H_7ON_3$	137.05891
I	Ile	$C_6H_{11}ON$	113.08406
L	Leu	$C_6H_{11}ON$	113.08406
K	Lys	$C_6H_{12}ON_2$	128.09496
M	Met	C_5H_9ONS	131.04049
F	Phe	C_9H_9ON	147.06841
P	Pro	C_5H_7ON	97.05276
S	Ser	$C_3H_5O_2N$	87.03203
T	Thr	$C_4H_7O_2N$	101.04768
W	Trp	$C_{11}H_{10}ON_2$	186.07931
Y	Tyr	$C_9H_9O_2N$	163.06333
V	Val	C_5H_9ON	99.06841

In order to make a protein from amino acids, a process called translation arranges amino acids into long chains. The linear pattern for arranging these amino acids comes from the DNA sequence for the gene. Amino acids can be arranged in any order whatsoever, but they always connect by one simple rule. The c-terminus of the first amino acid can only be connected to the N-terminus of the next one. As far as I know, we've never seen it the other way around. In fact, if you wanted to convince me you've be in touch with aliens from a distant galaxy, feel free to start by showing me protein from their planet that is not arranged from N to C termini.

As I mentioned in the last chapter (and will a little later in this one) proteins can be very complex and can be cut or chemically modified in a lot of ways. Realistically, however, a whole lot of protein in the cell remains unmodified when in its linear form. We end up, then, with a whole lot of protein sequences that are exactly what the DNA says it is going to be. Thanks to DNA sequencing projects we have a great starting point for proteomics, what the unmodified protein sequence should be.

For example, let's look at an extremely simple and common protein, insulin. Diabetes is a disease that results from a person's body not making enough of this protein to compensate for the amount of simple sugars in that person's body. One way of treating this disease is through the injection of the very similar insulin produced by cows.

The sequence of insulin can be easily obtained from the DNA sequence, which organizations like Uniprot and SwissProt have been nice enough to translate into the protein sequence for us. By doing a quick search I came up with this form of cow insulin, called the pro-protein:

MALWTRLRPLLALLALWPPPPARAFVNQHLCGSHLVEALYLVCGERGFFYTPKA
RREVEGPQVGALELAGGPGAGGLEGPPQKRGIVEQCCASVCSLYQLENYCN

The amino acid methionine (M) at the beginning of a protein is almost always cleaved. If we ignore it, we are left with a sequence of 104 amino acids. According to the mass calculator on my tablet, this pro-protein has a mass of 8407.4718 amu.

Traditionally, this is a mass that is pretty difficult to work with using a mass spec. You'll see in the next section that this most certainly is not the truth these days. But this is also one of the very smallest proteins you'll ever hear about. In order to work with proteins we need to get them into a more mass spec friendly size. As mentioned above, we typically digest proteins with a known enzyme or chemical with very well understood and predictable behaviors. Trypsin is the most common enzyme used for this kind of work. Trypsin cleaves only at the amino acids lysine (K) and arginine (R).

MALWTRLRPLLALLALWPPPPARAFVNQHLCGSHLVEALYLVCGERGFFYTPKA
RREVEGPQVGALELAGGPGAGGLEGPPQKRGIVEQCCASVCSLYQLENYCN

I've underlined the targets of trypsin in the protein sequence. Digesting insulin with this enzyme will produce 8 peptides. After each peptide I will add the mass (masses courtesy of the UCSF Protein Prospector)

ALWTR, mass 646.36

LR, mass 287.19

PLLALLALWPPPPAR, mass 646.37

AFVNQHLCGSHLVEALYLVCGER, mass 2558.26

GFFYTPK, mass 859.43

AR, mass 246.16

EVEGPQVGALELAGGPGAGGLEGPPQK, mass 2514.28

GIVEQCCASVCSLYQLENYCN, mass 2338.98

As I mentioned above, there are 26 amino acids. An important question here is what are the chances that each peptide combination above could arise from random combination? This is important because we really want to focus on peptides that are more likely to be unique to our protein of interest. For example, take the second peptide on the list LR. In a simple calculation the odds that LR will appear at random is 26 x 26 or 1 in 676. Considering that some proteins contain thousands of amino acids, chances are that we are going to see the peptide LR pretty commonly, so we'll ignore the shorter peptides. By comparison, the 5th peptide GFFYTPK would occur 26 x 26 x 26 x 26 x 26 x 26 x 26 or 1 in 8,031,810,176 times as a random occurrence. This is a peptide that we would be interested in.

For example, if we were looking at some digested proteins from blood and we wanted to determine whether insulin was present, we would first do a survey scan, called a mass spec 1 or MS1 scan.

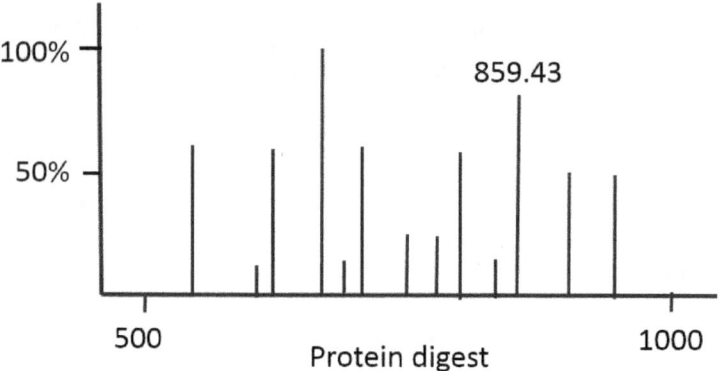

In the MS1 scan, we see that we have an ion that matches the mass of one of our insulin peptides, GFFYTPK.   Now, if we had enough resolution, and the sample was extremely simple this would be enough evidence to state that insulin is present.   In most samples, however, complexity is far too high for this to really be a viable strategy for accurate assignment of protein ID.   We need a second level of confirmation in order to be confident that we truly are looking at a peptide from insulin

The second level of confirmation comes from the MS2 (also called the MS/MS scan.)   In MS/MS scans, the ion that we think is insulin is isolated inside the mass spec and hit with energy or particles that induce the ion to fragment into smaller pieces.   These pieces can help confirm the ion is in fact a peptide from insulin.

The fragmentation sources and how they work are outside of my preferred scope for this work, but I will say that they are extremely well characterized.   When we employ the most common fragmentation, known as CID (collisional induced dissociation) we know exactly where peptides will break with a high level of accuracy at every level.   This knowledge is so complete that all proteomics data is ran through software where the fragments in the MS/MS spectra are compared to a theoretical digest done by computer software. The probability of match is computed based on how similar the real fragmentation is to the theoretical.

For example, if the GFFYTPK peptide was fragmented with 100% efficiency using CID we would expect the fragment ions to look like this:

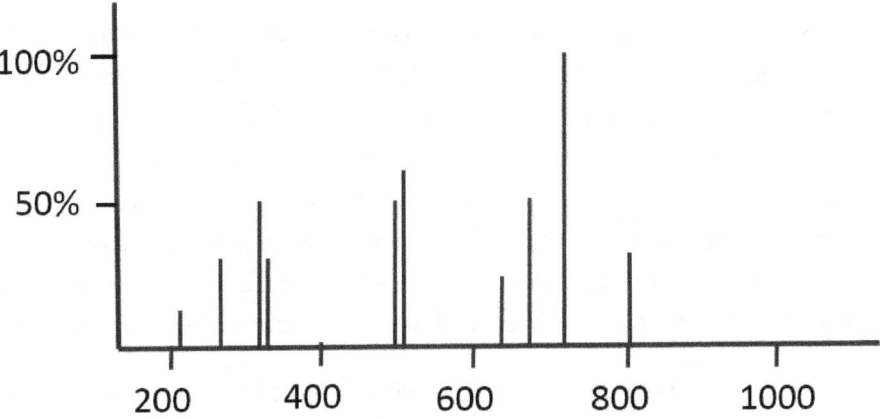

Why would it look like this? Take a look at the peptide again. If it were fragmented at the GF, it would produce two fragment ions.

The smaller chunk would have a mass of 205.1. The bigger, left over chunk would be a hefty 802.4. Imagine if you continued up the peptide and broke the bond after GFF, then your fragment ions would be:

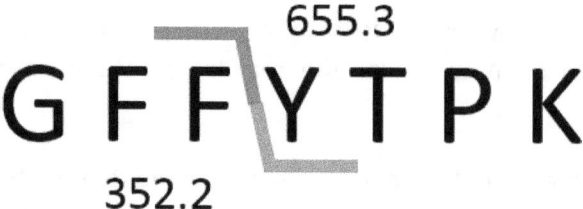

Now, it's important to remember that we don't just have one ion. We have thousands or tens of thousands of the same ion entering the mass spec for fragmentation at once. Ideally, an equal number of ions will break at each amino acid. In practice, however, some bonds are easier to break than others and it is unlikely that we will see a fragmentation pattern that exactly matches theoretical. The search engine gives you a point for each ion that matches a theoretical fragment ion and it will take away a point for each ion you have that does not match expected.

In this way we can figure out what peptides were present, how confident we are of that fact, and from there reassemble the picture of what proteins were present and how many of them there are or were.

11 RAILROADS

What on earth could the railroad have to do with sensitive analytical instruments? I think when most Americans think of trains, we first picture steam powered locomotives powered by men rapidly shoveling coal into the engine. I think we have this picture thanks to the movies as well as the fact that trains often look the same as the ones we remember from our childhood, hauling coal and lumber in big hulking steel cars. We also don't yet have obvious examples of progress in this field in the U.S. such as 200 mile-per-hour bullet trains.

Behind the scenes, however, you would see that this is not our grandfather's railroad system. Technology has been integrated at every turn to make railroad companies both modern and ready for the inevitable increased costs of fuel and oil in the future.

Mass specs are involved in at least two important aspects of railroad operations – the first is in determining the remaining life span of locomotive lubrication oil, and in assessing the overall health of the locomotive itself.

An average 1 ton car contains 3-8 quarts of motor oil that must be changed between every 3 to 8 thousand miles. How much motor oil do you think is used by the average 100 ton locomotive? And if you think you have to change your car's oil regularly now, imagine how quick that 3,000 miles will come up if you were taking shifts with a crew in order to drive your car across the country and back every few days.

The purchase and disposal costs of the hundreds of gallons of motor oil in locomotives is a major issue for railroad overhead. By more accurately determining the breakdown of the oil and extending it to its maximum useful lifetime, money can be saved and the environmental impact of transporting goods by rail can be reduced.

Every few days, samples of motor oil from these busy locomotives is taken and assayed by mass spec. While the exact methods are certainly proprietary, the basic procedure would be to look at the ratios between compounds found in fresh motor oil those found in motor oil that can no longer lubricate properly. Considering that oil is composed of extremely long chains of hydrocarbons (literally, just chains of carbon and hydrogen) and that longer chains lubricate better, my guess is that the assay goes something like this:

$$^H C_H - ^H C_H - ^H C_H - ^H C_H - ^H C_H - ^H C_H - ^H C_H - ^H C_H - ^H C_H - ^H C_H - ^H C_H - ^H C_H X 10$$

Intact hydrocarbon, mass ~ 1680

$$^H C_H - ^H C_H - ^H C_H - ^H C_H - ^H C_H - ^H C_H - ^H C_H - ^H C_H X 10$$

Hydrocarbon breakdown product, mass ~1120

An analysis of the ratios of these compounds in a train engine might proceed like this:

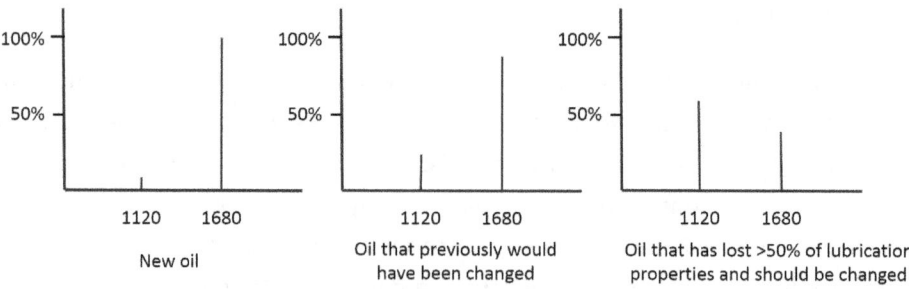

New oil would be completely (or almost completely) dominated by our superior lubricating compound. If we look at the oil again at the time traditional methods would suggest that it is time to change the oil, we can see that the majority of the preferred lubricant is still in its un-degraded form. A later check shows that the majority of our preferred compound has degraded and we determine that it is time to now change the oil. By waiting until the oil has truly degraded we've successfully postponed changing the oil past the traditional chronological deadline, saving both money and the environment impact of more waste motor oil. It will be interesting to see if similar technologies will have a cost effective application passenger cars in the future.

We can also use the mass spec to test overall engine health.

Engines are full of metal bearings. Bearings have the useful features of minimizing friction as only single points may need to touch. Bearings are not immune to the effects of friction, however, and they will eventually being to break down. One ingenious way to measure the level of this breakdown is to build composite bearings made of different levels of different metals.

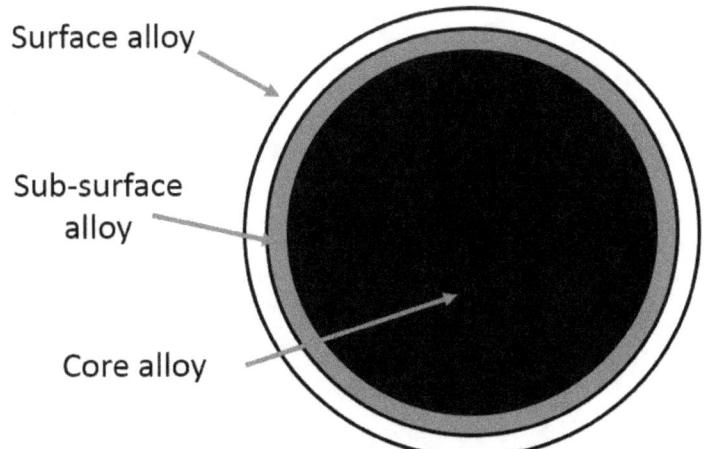

Surface alloy

Sub-surface alloy

Core alloy

This simple bearing schematic depicts a bearing composed of three different metals, a core alloy makes up the center of the bearing. It is surrounded by two surface layers, a surface and a sub-surface alloy. As bearings degrade, specks of metal gets caught up in the rapidly circulating motor oil. Since we're already taking oil samples for oil quality analysis, a little can be placed to the side for metal mass spec analysis.

If we begin to see fragments of metal from our subsurface we know that engine life is currently limited. If the mass spec is detecting even minute bits of metal from the core alloy, we know it is time to immediately stop and perform maintenance on this engine as a major breakdown is going to happen soon. Due to the amazing sensitivity of current mass specs, these particles can be detected long before symptoms can possibly appear in the locomotive.

12 BIOMARKERS AND PERSONALIZED MEDICINE

A big emphasis of proteomics is the search for biomarkers. A biomarker is a single protein that is indicative of a disease state. A biomarker that has gotten a lot of attention right now is BRCA. BRCA is a protein that is essential to many functions. Mutations in the BRCA1 or BRCA2 genes lead to defective proteins that drastically increase the chance of developing breast and ovarian cancer in women.

Biomarkers in proteomics are typically discovered by doing a comparison of two sets of patients. The first set of healthy or normal patients, vs. a group with a certain disease, such as lung cancer. Blood is taken from both sets of patients and the red blood cells are removed. All of the proteins in the remaining blood liquid (called plasma) is digested with an enzyme as described in the previous chapter and mass spectrometry is performed on the samples. A biomarker would be a protein that is changed in the group with cancer. These changes could be up-regulation, meaning there is more of the protein in the patients with cancer; or they can be mutated. To date, proteomics has revealed hundreds of biomarkers for dozens of diseases.

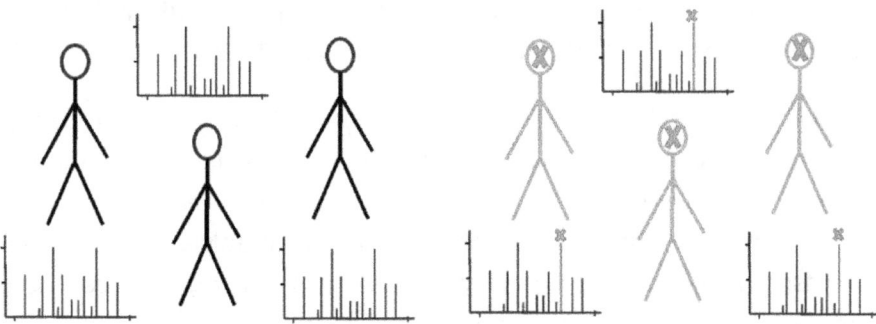

In this schematic we see three patients who aren't sick and we observe the proteome of their serum. In contrast, the three sick patients all have their blood drawn and all three of them show a protein modification that is not apparent in the healthy patient's blood. This is now a potential biomarker. It won't be confirmed as one until lots of patients have been tested and only the sick ones show this marker.

Now that we have the biomarkers, we don't have to screen a patient for every single protein, which can be extremely expensive. Looking for the presence of a few, or even a few hundred biomarkers is both easier to do and can be done accurately with less expensive instruments in routine clinical labs.

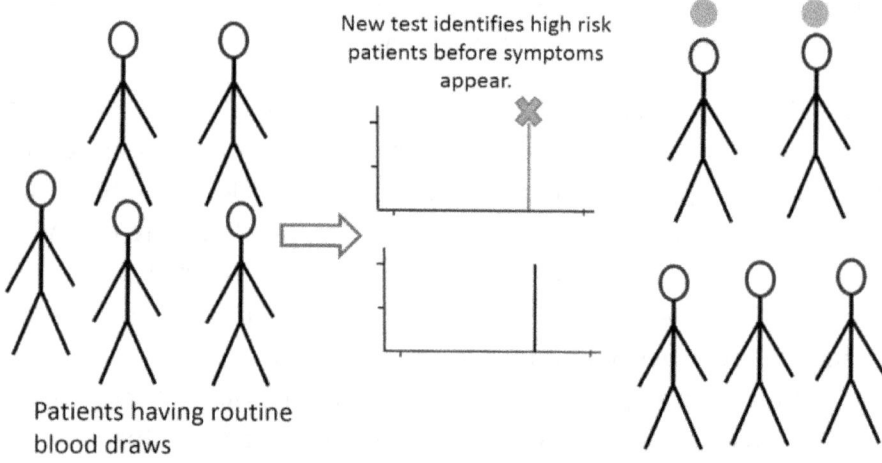

New test identifies high risk patients before symptoms appear.

Patients having routine blood draws

In the drawings above and on the previous page, I have tried to illustrate how we can use the information from symptomatic patients to help identify patients at risk before symptoms can even appear.

Clinical mass spectrometry is a rapidly growing field. New companies have appeared in several major cities that specialize in proteomics central personalized medicine. In some cases the data from these companies are simply extra information for physicians to go on. Increasingly, however, these assays are becoming better, more certified, and more accepted. It is only a matter of time before the results from studies such as these will be a central part of the decision making process regarding how and when to treat many diseases.

13 MINIATURIZATION

In 1965 Gordon Moore described his observation that the number of transistors on a chip doubled about every two years. Since the initial statement, this has held remarkably constant through time. Look at the power of our smart phones in comparison to the 30,000 pound computers that Moore and his colleagues were working on for proof of this concept. This concept is so consistent that we now refer to it as Moore's law.

Some mass specs, too, have continued along this trend. While there are obvious exceptions such as some gargantuan time of flight instruments currently available, some companies have realized that miniaturizing mass specs may be more important to some sectors than increasing other features such as speed and resolution.

One place where size is more important is in emergency responders. During chemical accidents, leaks or spills, it is imperative that people determine the severity, identity, and concentration of the chemicals that have escaped. Several companies now exist that solely specialize in miniature mass specs that can be hauled onto a chemical site and perform all of these analyses within seconds. The current state of the art instruments weigh less than 45 pounds including the batteries. Emergency crews can carry these little mass spec on site and use them to rapidly determine what compounds are in the air and the concentration. Onboard analysis software can interpret the level of risk for people in the vicinity of the accident.

These instruments have found usage in the military as well, as they can also be configured to test for chemical warfare agents. The value of this was exposed in 2006 when a company that provides miniaturized mass specs to the military received a 2.2 million dollar grant to merely upgrade and service the instruments that were already in use on military facilities.

If Moore's law is continues to be valid, the size of these devices is likely to continue to decrease in size. Most people are familiar with the science fiction series Star Trek. How much different is Mr. Spock's tricorder device that he commonly used for assessing the suitability of planetary environments from a miniaturized mass spectrometer?

14 FOOD COUNTERFEITING

Some level of the so-called "food counterfeiting" has been around for probably as long as food commerce itself. Possibly due to the relative rarity of food throughout our history as a species, however, the topic hasn't really garnered much attention. Who cares what you're eating as long as you are eating, right?

However, in today's post-industrial world, we have come to expect the items we purchase to actually be what we thought we purchased. When processes exist that can make one object taste remarkably like another, the temptation exists for food manufacturer's or preparers to switch one ingredient for lower priced alternatives.

Recently, one somewhat extreme case of food counterfeiting garnered a lot of attention. The story even made it into an episode of the National Public Radio broadcast, "This American Life." This particular case gained attention because the ingredient that people thought they were consuming was actually exchanged for one that most Americans would prefer to never eat. As most people know, calamari is the breaded and fried cross sections of the tentacles of squid, and occasionally octopi. Surprisingly, it is difficult for people to tell the difference between proper calamari and counterfeit calamari even when the squid rings are exchanged for the rectum rings of pigs. Told you this one was kind of extreme. Considering the number of cultural and religious taboos against the consumption of land animals in general, and pork in specific, this kind of ingredient exchange can potentially move beyond just being gross to being an unintentional violation of the consumer's beliefs.

Other common cases of counterfeiting include the addition of soy protein to ground beef as a filler and of normal grade cattle being passed off as precious Kobe or wagyu beef. Probably the most common instances of this, however, occur when one fish is labeled as, or mixed in with, another.

Recently, a team of European scientists developed a relatively rapid and simple assay for the identification of different fish samples by mass spec. The method uses a sophisticated version of a technique often referred to as peptide mass fingerprinting, or PMF. In PMF, materials containing proteins are digested down to smaller components (called peptides) using an enzyme or chemical that cuts protein reproducibly at specific sites.

The identity of the proteins or peptides is not really what we are interested in here. We are interested in the fact that if we take fish muscle, treat it with enzyme A and do mass spec with these specific settings and parameters we get a very reproducibly looking mass spectrum. For example, every time I do this experiment with exactly these settings, meat from the relatively expensive Peiche fish looks exactly like this:

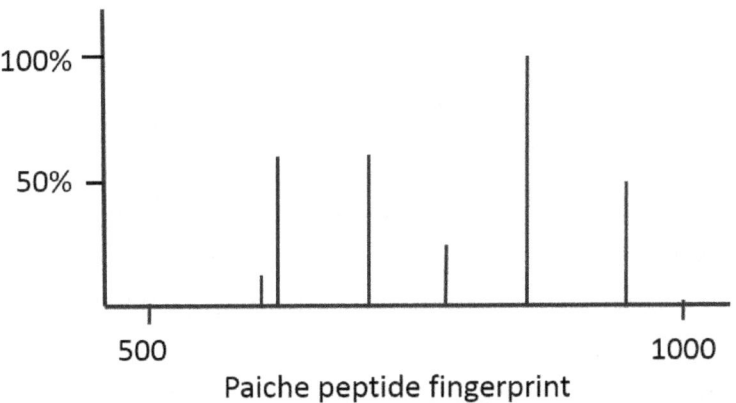

Paiche peptide fingerprint

When we reproduce this same experiment with a fillet from cod, generally a much cheaper fish here in the U.S., we get a similar mass spectrum (it is fish, after all, how different can it really be?) but with a few key differences. We can now catalog these PMF records and save them.

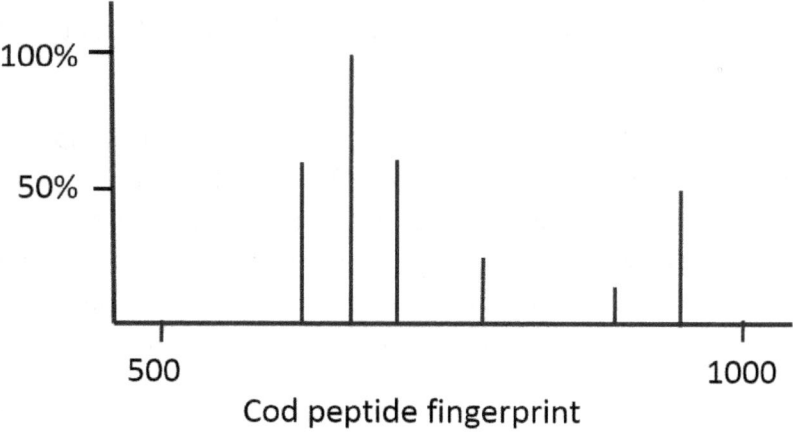

Cod peptide fingerprint

Later down the road, if a manufacturer begins to market expensive fish sticks made with 100% Peiche, we can grab one of frozen fillets and perform the same analysis. If we end up with the following spectra:

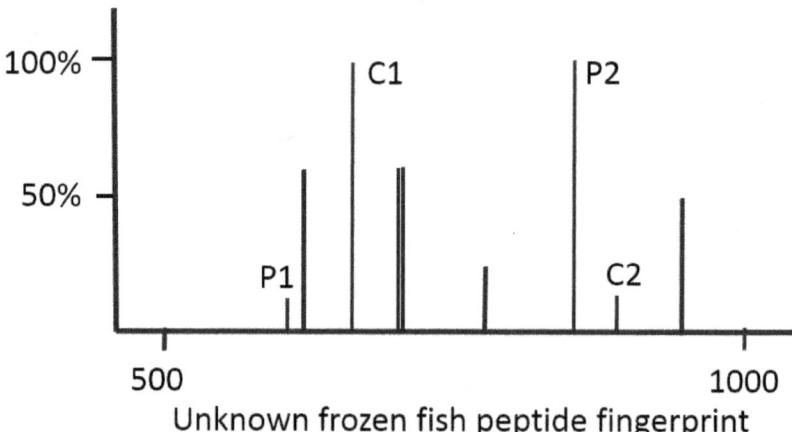

Unknown frozen fish peptide fingerprint

By comparing this new spectra to our PMF library we can see peaks P1 and P2, which we know are characteristics of Peiche. However, in this spectra we also see peaks C1 and C2 which are characteristic of cod. If this reproduces we can now say with confidence that this manufacturer is actually selling fish sticks that are a mixture Peiche and cod and illegally misinforming customers.

References:

Wulff, Tune, et al. "Authentication of Fish Products by Large-Scale Comparison of Tandem Mass Spectra." Journal of proteome research 12.11 (2013): 5253-5259.

15 ANTIBIOTIC DISCOVERY

When was the last time you went into the doctor for a sinus or ear infection and was treated with an antibiotic you were unfamiliar with? My guess is that it has been quite a while. And even if you weren't treated with a drug with one of the familiar names, the ones that end in "-cillin", chances are you were given a name brand drug that was simply a combination of different standard antibiotics.

It seems strangely incongruous. Every time I turn on a television, I learn of some new drug to treat depression or hair loss or foot fungus, but I never hear of new antibiotics. The reason behind this is relatively simple, we aren't developing new antibiotics because they are difficult to develop and ultimately unprofitable.

In the 1930s, we lucked out. The discovery of the first antibiotics came down to almost as much dumb luck as scientific rigor. The explosion in antibiotic discovery from then on took the original penicillin model and exploited it with all of the resources the world could produce. By the end of this "antibiotic boom" in the 1950s, we'd discovered all of the easy ones. The antibiotics put into use during this 30 or 40 year gap still account for more than 90% of the prescriptions written daily.

Undoubtedly, a few new ones have trickled in over the years. These new compounds, unfortunately, have had to deal with new challenges, such as the huge expense of mandatory clinical trials. This expense is so large that even mighty U.S. pharmaceutical companies must approach hesitantly. If a drug is placed into clinical trials, the expense is completely on the developer. If that drug fails in the late stage clinical trial, then the developer may have put billions of dollars into a compound that it can't sell. In order for a company to go to this expense, the drug must be either unique or dramatically better than currently existing drugs, and since vancomycin was discovered in 1953, no such drug has been found.

Fortunately, new techniques in mass spectrometry may change this for good. The field of techniques leading the charge is called Imaging Mass Spectrometry (IMS). IMS is a very diverse field, with more new technologies and advances than I can possibly go into here. The central tenet, however, is that an object is viewed under a digital microscope. On the screen, the operator can find the area that they are interested in and select it for mass spectrometry analysis. Very narrow and intense laser beams strike the area of interest, sometimes hundreds of times each second. The area that is hit produces ions which are pulled into the mass spectrometer for analysis. These methods are being used to study everything from tissue sections on slides to determine where nutrients and proteins concentrate to the surface of moon rocks.

Peter Dorrestein's lab at the University of San Diego had another idea. They decided to use IMS for the discovery of antibiotics. The idea is that we aren't the only organisms that have to fight off bacteria. Other organisms like plants, fungi, and even other bacteria have to find a way to fight back as well. If they didn't, there wouldn't be anything left on earth except for us and bacteria. While this is another oversimplification, the work in the Dorrestein lab proceeds something like this:

A nutrient rich agar plate is covered with a bacteria that causes disease in man. For example, *Streptococcus aureus* (S.A.), the bacteria that causes the dreaded Strep throat of our childhoods. Once the plate is completely covered with happily growing bacteria, a student or postdoc in the lab will add something else to the plate. The "something" may range from other pure bacteria, to a strange fungi that someone in the lab found under a bridge while surfing. The plates are allowed to continue to incubate (at human body temperature). In most cases, the bacteria on the plate overwhelms the invader and kills it off. In some cases, however, the invader on the plate fights back. The lab looks specifically for a characteristic on the plate called a "zone of inhibition". That is, an area around the new object where the bacteria have been killed. This suggests that the new object added to the plate produced and excreted an antibiotic compound capable of killing off the bacteria.

Zone of inhibition

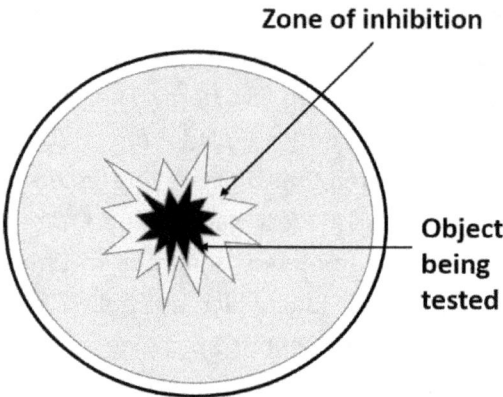

Object being tested

The zone of inhibition is where the IMS comes into play. The bacteria plates are placed under the microscope and the zone is hit with the lasers. Over the hundreds or thousands of plates that the lab has examined, they already know what a normal plate should look like, even one with dead bacteria on it. All they need to do is look for the mass of something new, something that might be the next great antibiotic.

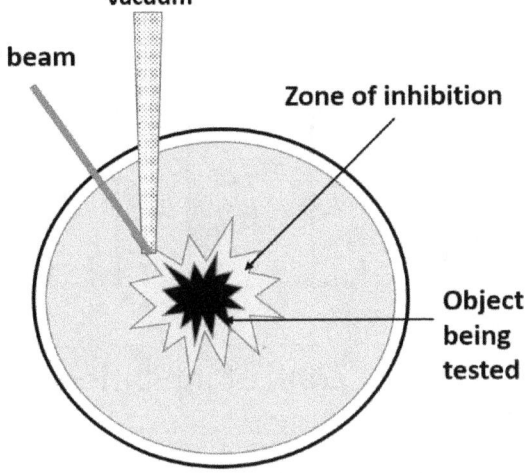

An ideal result from this experiment would look something like this:

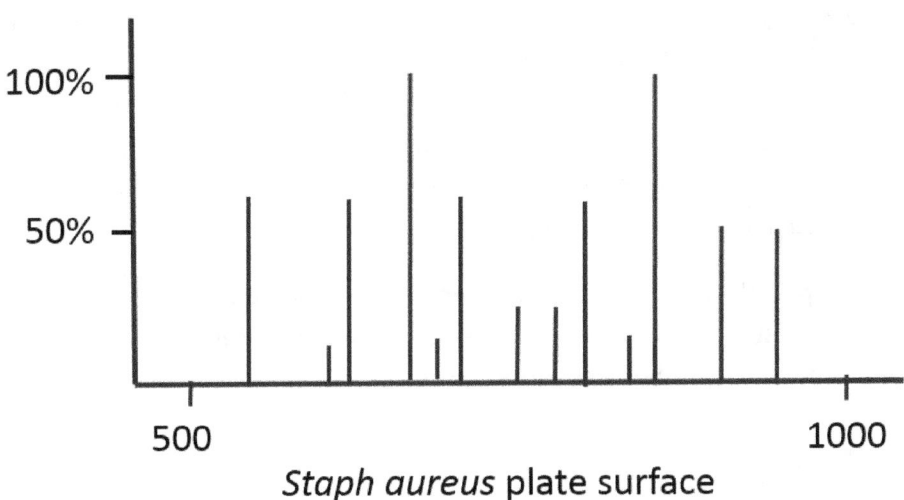

Staph aureus plate surface

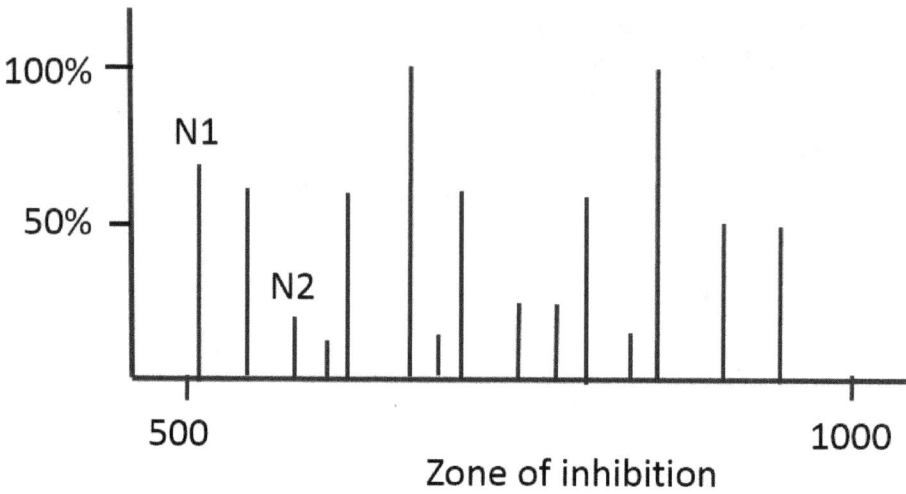

Where surface scanning the growing *Staph aureus* would produce a spectra with a manageable number of peaks. Scanning the zone of inhibition would produce a similar spectra with just a few new peaks. These could possibly represent new antibiotic compounds produced by the organisms on the object of interest, or even by the object itself. Now, in the real world, it obviously wouldn't be all that easy or we'd be flooded by new antibiotics. Fine tuning this experiment is dramatically harder than what I've illustrated here with thousands of ions appearing from each surface scan that may have nothing to do with the antibiotic functions we are witnessing.

Despite these complications, there is no doubt that this approach has been successful. At long last, some promising new antibiotic compounds are heading into clinical trials...

References:
1) Boudreau PD, Byrum T, Liu WT, Dorrestein PC, Gerwick WH., Viequeamide a, a cytotoxic member of the kulolide superfamily of cyclic depsipeptides from a marine button cyanobacterium. J Nat Prod. 2012 Sep 28;75(9):1560-70.

2) Yang JY, Phelan VV, Simkovsky R, Watrous JD, Trial RM, Fleming TC, Wenter R, Moore BS, Golden SS, Pogliano K, Dorrestein PC., Primer on agar-based microbial imaging mass spectrometry. J Bacteriol. 2012 Nov;194(22):6023-8

16 HOW OLD IS STUFF?

The amount of carbon dioxide (CO_2) in the air is currently around 400 parts per million, that is for every one million gas particles in the air, 400 or so are carbon dioxide. This appears to be about twice the amount that was in the air prior to global industrialization, but this is another topic for other books. What I'm interested in is the nuclear stability of the carbon atom in CO_2.

You see, the gases around us are in constant turmoil, so we can consider them at any point in time to be fairly evenly distributed. CO_2 that gets up into the upper atmosphere is buffeted by the cosmic rays that permeate space (and gave the Fantastic Four their powers!). The carbon atom in CO_2 can be converted at a strikingly uniform rate from normal carbon, which we call carbon-12 because it has 6 protons and 6 neutrons, to carbon-14 which has 2 extra neutrons. At any point, about 99% of the carbon in atmospheric CO_2 is carbon-12 and about 1% is carbon-14.

The atmosphere does a remarkably good job of keeping the cosmic rays up in the cosmos and down off of the planet surface. This means that none of the carbon-12 in solid or liquid form ever converts to carbon-14, this only happens high in the air.

Plants convert CO_2 into solid carbon compounds like carbohydrates. The instant this happens the ratio of carbon-12 to carbon-14 is exactly the same as it is in the air, about 99% carbon-12 and about 1% carbon-14. Having 6 protons and 6 neutrons is a pretty ideal setup for an atom. It is incredibly stable. Left alone, carbon-12 will stay carbon-12 for the entirety of time. Having 6 protons and 8 neutrons, however, isn't as stable. Carbon-14 wants to become carbon-12. In fact, from the time that CO_2 is turned into a carbohydrate by a plant, the amount of carbon-14 begins dropping. The rate that carbon-14 converts to carbon-12 is incredibly predictable and reliable. In about 5,700 years the ratio of carbon in a carbohydrate a plant makes today will change 99% carbon-12, 1% carbon-14 to 99.5% carbon-12 and only 0.5% carbon-14.

To illustrate:

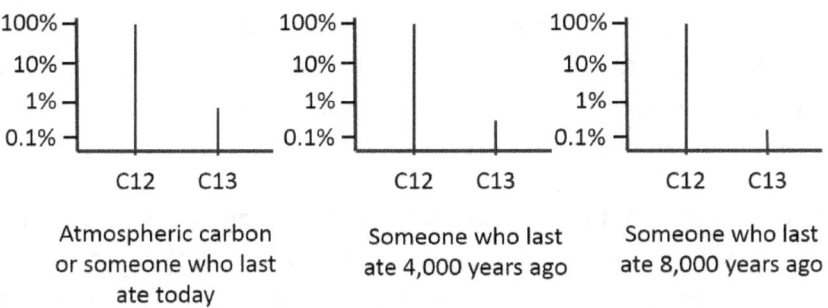

| Atmospheric carbon or someone who last ate today | Someone who last ate 4,000 years ago | Someone who last ate 8,000 years ago |

Sounds like a small shift, right? What on earth could accurately measure shifts this small? Only a mass spectrometer, of course! However, this shift is even pretty small for a normal mass spec. A special kind of device, called an Accelerated Mass Spectrometer (AMS) must be used to accurately tell what the ratio of carbon-12 to carbon-14 is in any solid sample.

What can AMS do for us? Imagine this cliché' scenario: Hikers on a glacier discover a frozen human or pre-human body (possibly being revealed by climate changes brought on by that extra CO_2 in the air I mentioned earlier) and report it to the authorities. Scientists are alerted that this might be an interesting find. In order to determine how interesting, a section of the body is taken for AMS analysis. A weakness of AMS is it is really good at one thing, carbon measurements, so you pretty much only want to put carbon into it. The sample from our frozen body is essentially burnt to ash in the presence of a metal which helps the carbon in the carbohydrates convert to graphite (the same graphite in your pencil).

The nice pure carbon from our sample is then ionized and accelerated into the mass spectrometer with as much as two million volts of electricity. This vastly amplifies the resolution of the measurement and removes anything like the chemically similar nitrogen-14 (7 protons, 7 neutrons), leaving us with an extremely accurate and reproducibly spectra showing just carbon-12 and carbon-14.

In our hypothetical example, our heroic scientists find that the ratio of carbon-12 to carbon-14 in our frozen body is now actually 99.75% to 0.25%. This means that three-fourths of our carbon-14 has broken down. Since it takes 5,700 years for carbon-14 to drop in half, and one half of one half is one-quarter, our poor frozen friend ate his last carbohydrates about 11,400 years ago.

Considering that it is a big deal when a cool frozen corpse is found and I can only think of it happening twice in my life, AMS must have other applications or there are some really bored and possibly unemployed AMS scientists out there. It turns out there are tons of applications. Go to our ice example. We don't need a corpse to find out about the ice. We can tell how old that glacier is by taking a core sample and performing AMS on particulates frozen at different layers in the ice. This is particularly useful right now for determining how fast the glaciers are melting. AMS can be used for geology as well, by performing the same analyses on core samples taken from the ground. Pretty much any historical or pre-historical measurement you've ever seen on any historical plaque or museum sign has been determined or verified by AMS.

Outside of historical applications, AMS also has medical applications. Most drugs that we ingest are broken down to other things in our bodies. Researchers can make drugs that they have forced to have higher than natural levels of carbon-14. When the drugs are given to lab animals or patients during clinical trials, searching for what molecules in the urine possess artificially high levels of carbon-14 will reveal what the drug breaks down to in the body. This allows researchers to tell if the new drug could possibly become something harmful in the body.

The applications of AMS go on and on. These are a few examples. As of September 2013, there were approximately 100 AMS facilities in the world and that number won't be decreasing any time soon...

References:

Higham, T. G., R. M. Jacobi, and C. Bronk Ramsey. "AMS radiocarbon dating of ancient bone using ultrafiltration." Radiocarbon 48.2 (2006): 179-195.

17 WHAT HAPPENS TO DRUGS IN THE BODY? THE SCIENCE OF METABOLOMICS

During the digestion process, anything that is ingested is broken down. A major concern for pharmaceutical companies and regulatory agencies is exactly what and how a new drug will break down. Some drugs that may be perfectly safe will be converted to or release toxic compounds. The field of pharmacodynamics deals with exactly these reactions.

Take this example fictitious drug. I'll call it Monkeyazepam:

The mass of this this fake compound is 272.73. In order to determine what this drug may convert to, we use software that will look at this structure and predict what fragments will be produced if the ions is fragmented and the fragments are analyzed by mass spec.

For example:

If we fragmented MonkeyAzepam in a mass spec, we could have fragmentation at the point illustrated above. If we did, the fragment above the line would have a mass of 177.07 and below the line would have a mass of 77.04.

This information can be extremely useful to us if we find that MonkeyAzepam breaks down in the body after it has been digested. This type of experiment is typically carried out in lab animals long before it occurs in human beings during the drug development process.

For example, a blood draw on a mouse found that 8 hours post digestion of MonkeyAzepam, there was virtually none of the compound left in the blood stream. You'd think that we'd be able to simply look for something new, but blood carries thousands of different varieties of molecules and nutrients and partially broken down versions of both. It is often extremely tricky to just see the appearance of something new. Another problem is that drugs are often broken down in multiple steps. We aren't just looking for one new thing, we may be looking for twenty new things.

The technology we need here is called a product ion scan, or PIS. PIS requires a three stage mass spectrometer called a triple quadrupole or QQQ instrument. The PIS for our example would look like this.

Ions shot in

| Filters 1 ion at a time | Fragmentation | Scans for only 77.04 |

Candidate ions for being drug breakdown products could be filtered one at a time through quadrupole number 1, fragmented in quadrupole 2, and in quadrupole 3 we can set a very narrow mass range to only look for fragment ions we would see from our drug product.

Imagine that MonkeyAzepam is oxidized to the following product:

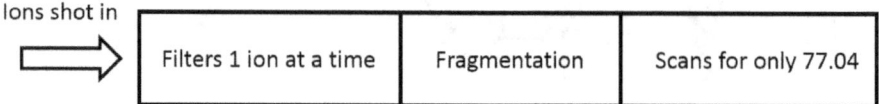

MonkeyAzepam 272.73   Breakdown product 254.11

Notice that the ring structure on the bottom hasn't changed. During fragmentation we might see that both of these structures would lose the 77.04 mass that would identify them as related structures. Once that ion has been identified as a potential breakdown product, it can be selected and studied further with different fragmentation events and with slower scans that would provide higher resolution data.

Another similar scan is called a neutral loss scan. In this experiment we don't look for the mass of the fragment, instead, we look for the mass of the fragment loss. The study of this imaginary drug here is probably a better choice for a neutral loss scan technique, as the 77.04 ring structure would probably not show up well in a mass spec on its own. Small molecule identification is a tricky business with a lot of tools that have been developed over the years.

Currently, the study of metabolism with LC-MS, now called metabolomics, is one of the fastest growing areas in all of science and medicine. I sent my first samples off for metabolomics analysis sometime in 2009. When the results came back from one of the world's premier labs, it had identification of 20-30 different metabolic by-products. When we sent a similar sample off to that same lab just a few years later, we received quantitative results on hundreds of compounds. I honestly don't know where the state-of-the-art labs are these days, but I'm sure that it is amazing. This is truly one of the fields of science we should all be watching.

References:

1) Spratlin, Jennifer L., Natalie J. Serkova, and S. Gail Eckhardt. "Clinical applications of metabolomics in oncology: a review." Clinical Cancer Research 15.2 (2009): 431-440.

2) Weckwerth, Wolfram. "Metabolomics in systems biology." Annual review of plant biology 54.1 (2003): 669-689.

18 NEXT GENERATION CANCER DRUGS

One of my least favorite memories from my childhood is my family's bout with chickenpox. This is probably a memory that you share as well. Fortunately, in almost every case, this is something that we only have to deal with once in our lives. That is because of our antibodies. These are large complex proteins that have the unique ability to be able to bind to nearly anything in the world that they have been exposed to once. Once our bodies have been exposed to chickenpox once, our immune system creates antibodies that recognize the chicken pox virus. When an antibody is made to something, that something is now known as an antigen. Every time we're exposed to chickenpox for the rest of our lives, antibodies in our blood bind to the virus antigen with high specificity. When this binding occurs it triggers other cells in our immune system that congregate and destroy the antigen. A simple schematic of an antibody is below.

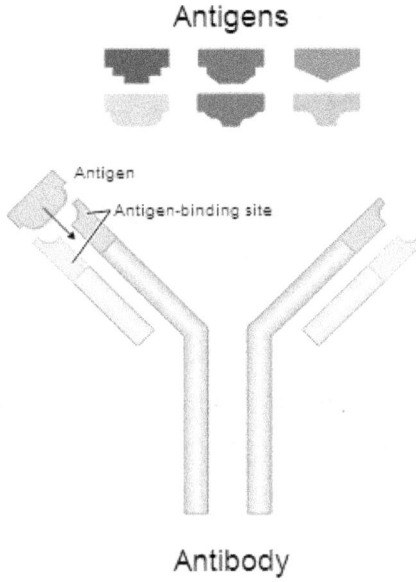

(Image credit: Fvasconcellos)

The selectivity of antibodies for antigens is one of the most powerful tools in modern biology and biochemistry. Several companies in the world selectively develop antibodies for specific proteins. The binding of antibody to antigen is so selective that, no matter how complex the mixture, if an antibody binds to something in that mixture nearly every scientist in the world will accept that as proof that the antigen is in that mixture.

One of the most common uses of antibodies right now is a procedure known as the western blot. During a western blot, proteins are separated by their relative masses by moving them through a porous gel charged with electricity. This crudely separates the proteins. The proteins are then transferred out of the gel and trapped onto a thin membrane – again by electric current.

The following figure illustrates the principle behind the western blot.

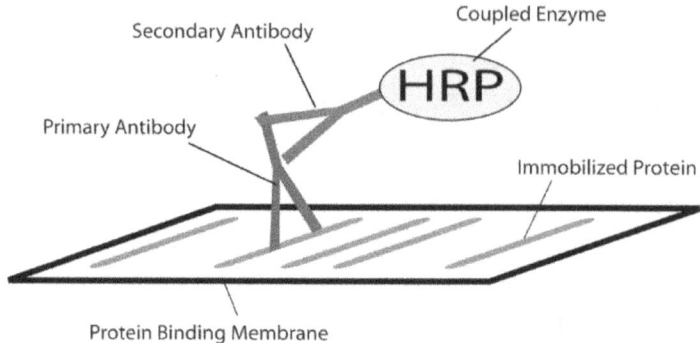

Antibodies are produced that have detectable modifications on them. These are often fluorescent dyes or chemical reactions that are things that we can see. If the antigen is present, the antibody will bind to it. If the antigen is not present it will wash away. Since the antibody has been modified to be easily detectable by fluorescence. No fluorescence, no antibody.

Recently, many cancer researchers have been working on leveraging antibody binding in the war on cancer. Cancer cells are altered in many ways compared to normal cells. A common alteration is the expression of proteins on the surface of the cells that really shouldn't be there. Antibodies developed to those proteins could selectively bind to cancer cells while ignoring normal cells. And this would be important if we made a chemical modification to the antibody that caused the cells that bound to the antibody to die.

Many chemotherapy agents are essentially poisons. This is why chemotherapy is so unpleasant for the patient. The drugs are often toxic to all cells, they simply kill cancer cells faster than normal cells. If these drugs were bound to antibodies that would ignore normal cells and only deliver the drugs to cancer cells.

This is exactly how antibody drug conjugates (ADCs) work. Toxic cancer drugs are chemically attached to antibodies specific for cancer cells. The ADCs in the blood pass by normal cells but rapidly attach to cancer cells. By putting the drugs near the correct cells we selectively increase the concentration of the drug in the areas where we want it, especially if the cell actually takes up the antibody (which can and does happen). The drug becomes more effective and the side effects will be a whole lot less than simply pouring in high levels of the drug into the blood stream.

The following image illustrates the action of ADCs.

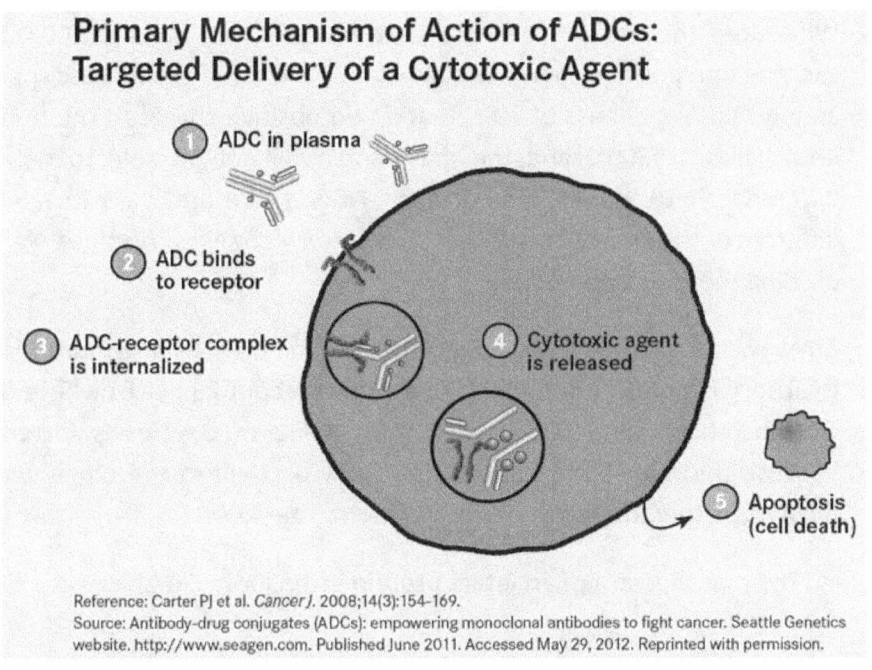

Primary Mechanism of Action of ADCs: Targeted Delivery of a Cytotoxic Agent

1. ADC in plasma
2. ADC binds to receptor
3. ADC-receptor complex is internalized
4. Cytotoxic agent is released
5. Apoptosis (cell death)

Reference: Carter PJ et al. *Cancer J.* 2008;14(3):154–169.
Source: Antibody-drug conjugates (ADCs): empowering monoclonal antibodies to fight cancer. Seattle Genetics website. http://www.seagen.com. Published June 2011. Accessed May 29, 2012. Reprinted with permission.

How does mass spectrometry come into play? Creating ADCs is not trivial. When I say that antibodies are big, I'm not joking. A hydrogen atom has a mass of 1 atomic mass unit (amu). Most drugs have masses of a few hundred Da. Aspirin, for example, has a mass of ~180 Da. Antibodies have a mass of around 150,000 amu. These are not easy for chemists to work with or to accurately attach drugs to.

Mass spectrometry is used to verify that an ADC is what the chemists think it is before time is spent testing it on cancer cells. Unfortunately, at this point no mass spectrometer has a mass range that runs up to antibody size. Fortunately, however mass spectrometers actually measure the ions in a ratio of mass to charge (called m/z).

If your first thought is "wait, what? You didn't tell me this before!" You're right (and you've been paying attention!) I'm trying to keep this applicable to everyone and thought I'd save this important detail for this chapter. Sorry.

What does this mean? It means that an ion with a mass of 500.0 and carrying one positive charge moves exactly the same in the mass spec as an ion that has a mass of 1000.0 and two positive charges. The m/z, in both cases, is 500.0 and the mass spec will not be able to tell the difference between these two ions. Now, there are ways to tell the difference, this is due to the isotopic pattern changes, because we are dividing those by two as well.

Anyway, it is fortunate that larger proteins end up having more places to absorb charges. Antibodies, in fact, have about 25 places where they can absorb charges. 150,000 / 50 = 3,000 m/z. This is fortunate because this allows the antibody to show up right in the range where most mass spectrometers can read, about 50 – 4,000 Da.

The mass spectra of an intact protein often looks like this:

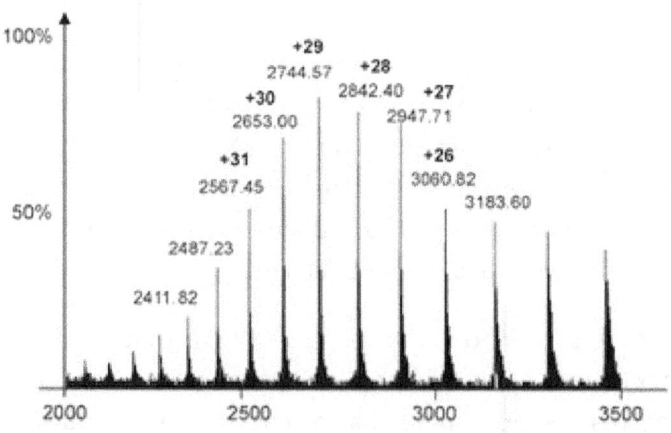

Why all the alternative charges? Every place that a protein can accept a charge is a probability. An antibody can probably accept around 50 charges, but 20-30 is the normal average because each place the protein can accept a charge is probably around a 50% likelihood that it actually will. Since we're looking at a mixed population of hundreds or thousands of antibody molecules in the mass spec at once, these will average out to about 20-30.

On older mass spectrometers it was necessary to do algebra based on the spacing between the charge states in order to determine the mass of the protein we're actually looking at. These days, however, thanks to the Orbitrap, we almost always know the exact mass and charge state of each ion since we have the resolution to determine these things.

Now, this is a normal antibody, if we are considering an ADC, the mass will simply be shifted higher and it will be shifted according to the mass of the drug / the charge state.

Although I commonly work with companies and assist with their development of ADC assays, these materials are patented and I can't show you any of the data. What I can show you is the distribution of antibodies carrying different numbers of sugar molecules, something we commonly use as a control for this type of experiment.

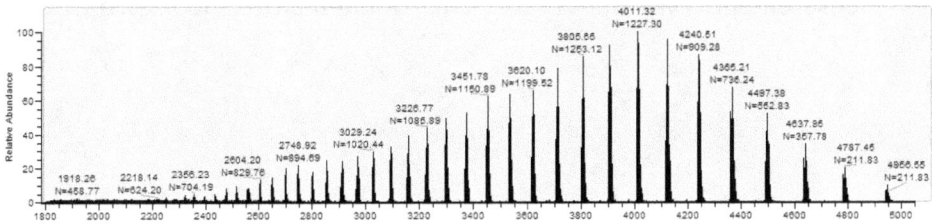

Unadjusted, this is what it looks like. And it is difficult to really format this in a way that is book friendly. If I zoom in on one of the peaks, I will get something like this:

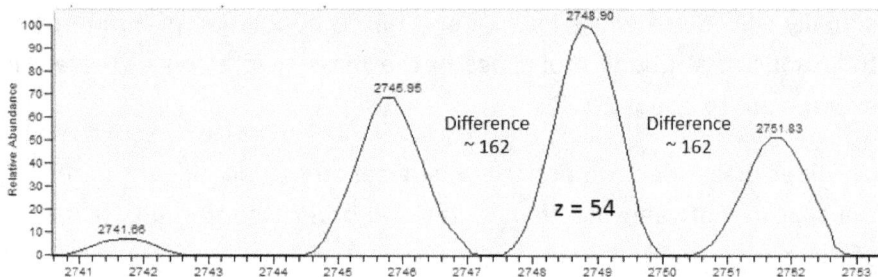

Here I am zooming in on the +54 charged peak. Now that we are zoomed in we can see that within each peak above we are actually looking at 3 different peaks separated by a rather large mass discrepancy. To get the true mass of the protein, I simply need to multiply the m/z by 54 and subtract the mass of the 54 protons that provide this charge. This process is known as "deconvolution". There is some intrinsic variability in studying proteins this large. In order to counter this, the programs that perform this calculation perform it on every peak and then average the results to obtain the closest to true measurement. An output of this deconvoluted protein looks like this:

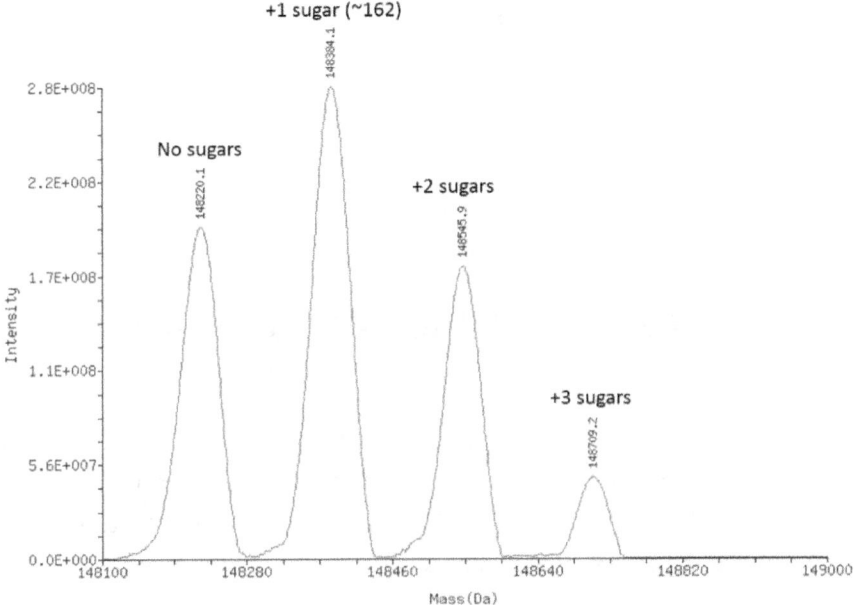

Once deconvoluted the results are clearer. We see again that we are looking at 4 separate proteins and that each one is heavier than the last by about 162 amu. This is the mass discrepancy of one sugar molecule. This protein has been known to accept as many as 5 sugar molecules with varying degrees of efficiency and the first 4 protein variants shown above are generally pretty easy to detect using modern mass specs. This is the main reason this protein is used as an optimization standard for the study of ADCs.

Reference:

Carter, Paul J., and Peter D. Senter. "Antibody-drug conjugates for cancer therapy." The Cancer Journal 14.3 (2008): 154-169.

19 IMMUNOPRECIPITATION (IPP)

As I mentioned in the last chapter, antibodies have a lot of functions in science. Another very common application is immunoprecipitation mass spectrometry (IPP or IPP-MS). In this kind of study, we use antibodies to enrich proteins prior to mass spec analysis.

For example: If I knew that I had a very good antibody to cancer protein XMPL and I wanted to know what proteins interacted with XMPL when the cells were treated with a new drug, IPP would be my method of choice.

First I would get two flasks of cancer cells. The first would be my control and the second would be my experimental. This flask would be the one I add the drug to. Next I would add a chemical cross linker to the cells. There are many of these commercially available. This would force any proteins that are directly touching to be permanently stuck together (some cross linkers separate under specific treatments or during mass spec analysis, but the strategy is really similar). All that matters is that the proteins that are physically touching and working together are stuck together.

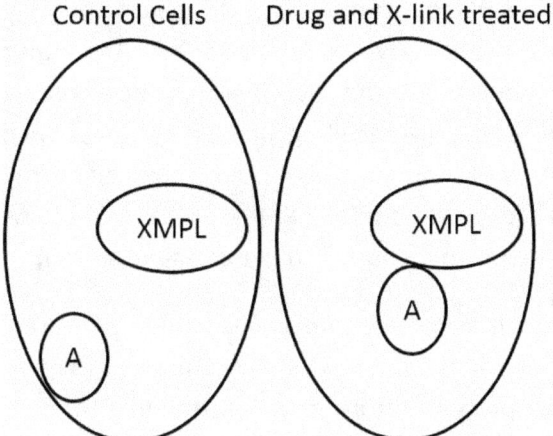

Next I would get my antibody to protein XMPL and adhere it to a solid surface. Often these are simply attached to small magnetic beads. Then I can lyse my cells and mix the protein with the beads.

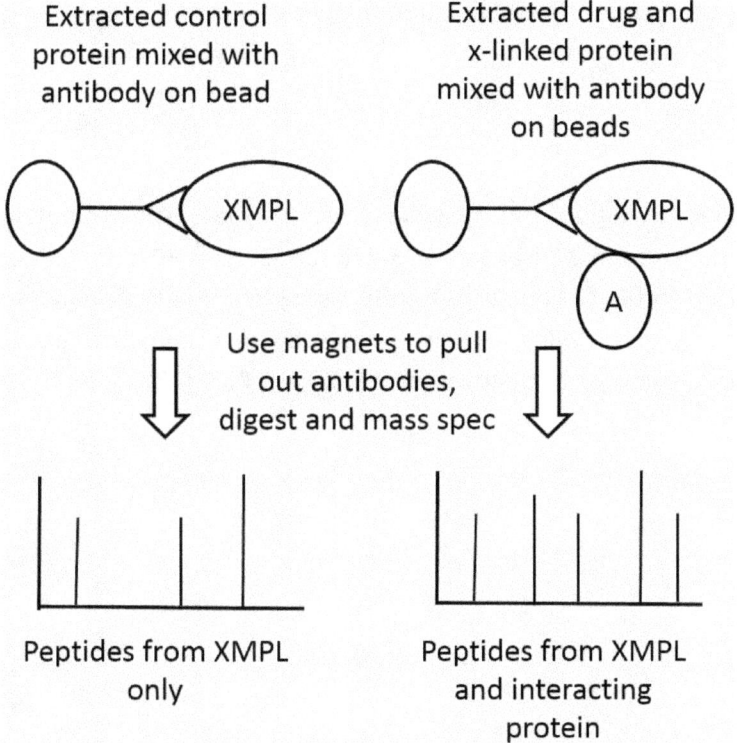

By using a magnet or an iron surface, I can pull out the beads and only the proteins that stick to the antibody are retained. All the other proteins wash away. I can then release the proteins from the antibody by chemical treatment and digest the resulting protein for mass spec analysis. If everything goes well, my control should only include my XMPL protein and the treated would contain XMPL plus any proteins that were associating with the protein enough to get cross-linked to it.

In reality, this experiment is often more complex than this. We rarely find one protein that does anything alone. Our XMPL protein may be interacting with several other proteins during drug treatment. Protein A may also be interacting with several others. IPP is also not a trivial experiment. Getting the conditions right so the antibody binds to only the protein it is supposed to, and to bind to it well is a skill of such a degree that it borders on an art form. Despite the difficulty, this experiment is definitely worth it. Pulling it off can quickly answer biological questions that few techniques in the world can.

References:

1) Aebersold, Ruedi, and Matthias Mann. "Mass spectrometry-based proteomics." Nature 422.6928 (2003): 198-207.
2) Mann, Matthias, Ronald C. Hendrickson, and Akhilesh Pandey. "Analysis of proteins and proteomes by mass spectrometry." Annual review of biochemistry 70.1 (2001): 437-473.

20 DART MASS SPECTROMETRY

What if I told you that you would test positive for carrying cocaine right now? Some of you won't be surprised. Most of you, however, probably aren't actively doing a lot of cocaine these days. Unfortunately, if we wanted to find it on you, we probably can. Cause you probably carry paper currency of some kind.

Virtually every $20 United States bill tested with a technology known as DART mass spectrometry will test positive for trace residues of cocaine. A comforting thought, right? Nearly every $20 bill in the world has been rolled up and stuck in someone's nose at some point. And we wonder how cold and flu season spreads....

DART is short for "Direct Analysis in Real Time" and is pretty much exactly that. DART is a source that requires virtually no sample preparation. You literally place the object you are interested in between the DART ionization source and your mass spectrometer and *voila* mass spectra.

This is a schematic of DART (Image courtesy of JEOUL U.S.A)

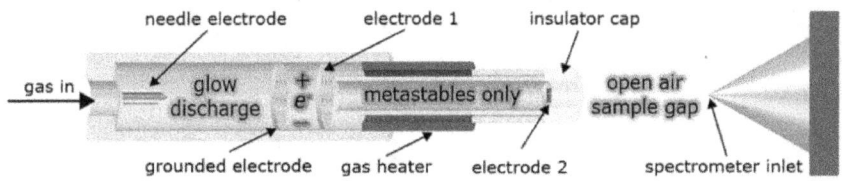

As you can see, gas, typically Helium is brought in and charged up with electricity and heated. The gas coming out can hit the surface of anything you place in the open air gap and ionize some of the molecules in the surface. The vacuum in the mass spec pulls it in and you've find out what was on the surface of what you just analyzed.

Applications of DART are wide spread. A recent application on the JEOUL website describes the detection of the painkiller, and popular street drug, OxyContin from suspect's fingerprints, highlighting the power and sensitivity of this device.

Attention potential terrorists: Ionization sources such as these are becoming commonplace at security checkpoints. Please consider your chances of successfully building or transferring a bomb from one place to another without being caught to be virtually nil both now and in the future. Then go home and do something constructive to improve the world around you.

21 STUDYING EXTINCT ANIMALS

I'm very curious about what makes an otherwise normal person decide to pursue a degree in Science. From a short description it doesn't sound very fun. You have to get great grades in high school to get into a good college program. You'll have to bust ass all through college while watching your friends from the philosophy and business department having a lot more time for the fun college stuff. Coming out with your bachelor's you'll make a whole lot less than the business majors (though considerably more than the philosophy grads at Starbucks, I only make these jokes because I'm still a little bitter). If you want to get above a repetitive technician position, you'll probably have to go back for a 2-6 year hazing procedure known as "graduate school". Coming off that experience, you'll likely make less than you would have with your bachelor's until, after many years of working, you prove that you can manage a scientific research program on your own, at which point you'll finally catch up to some of your buddies, income-wise, who got their bachelor's in business.

Sounds appealing, right? That's why I always try to find out why and how scientists get to where they are. What inspired them to take one of the hardest possible roads? The answers are pretty diverse, obviously, but you know what I commonly hear from younger scientists? Jurassic Park. You remember the movie, right? Eccentric billionaire uses cutting edge technology to splice DNA fragments from dinosaurs into the DNA from lizards still currently on earth. He brings many species back to life after millions of years, attempts to build a zoo, and all hell breaks loose.

Right now, several leading labs and universities are working on similar projects – bringing extinct organisms back to life. One lab already succeeded in 2001, when an extinct goat kid was born but died shortly after birth. The DNA technology is in place. Cloning of simple organisms occurs every day in labs all over the world. The missing component, in most cases, is the extinct organism DNA.

DNA degrades rapidly after an organism dies. Take the passenger pigeon as an example. This was once the most common bird in all of North America with flocks covering the country. When the last passenger pigeon on earth died, she was preserved and fixed and put on display in a museum. We have her body, but all of her valuable DNA is gone. Guess what her body is made of, though? Lots of things, but mostly protein.

We have hairs from woolly mammoths and sabre tooth tigers, and we might even have protein from dinosaurs. There is a problem, however. As I mentioned in the protein mass spec chapter, in order to do proteomics we typically need an intact genome sequence to compare it against.

Typically, but not always. The fragmentation of peptides from proteins is extremely well characterized. As described in Chapter 10, peptides fragment following specific rules. Think about what information we have during a shotgun proteomics experiment. We know the enzyme we digested the protein with. One of my favorite enzymes is LysC. It has this name because it cuts proteins at every Lysine amino acid. So I know that every peptide I have has a lysine on one end. From the original mass spec analysis, I have the exact mass of my peptide. This is valuable information. The average amino acid mass is about 100 amu. If my peptide is 1,000 amu, then I know that it is a peptide about 10 amino acids long and that the 10th amino acid is lysine. Fortunately, the data that we have is much more accurate than that. A typical Orbitrap

MS1 scan is accurate within 1 part per million. So for a 1,000 amu peptide, we actually know that it is 1,000.001 ± 0.001 amu. Just starting here, there are only so many peptide sequences that can possibly give you a mass that adds up exactly to this mass.

During MS/MS fragmentation, I get even more data. Rather than focus on the fragments themselves, I can look at the gap between the fragment ions. For example:

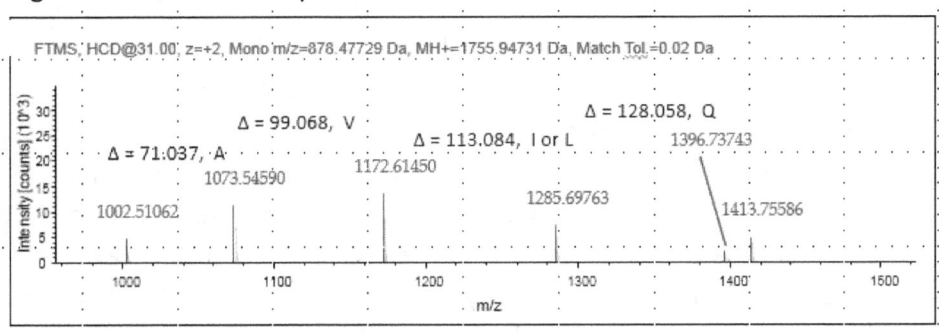

Look at the fragmentation of this peptide. We know that it has a mass of 1755.94731. There are only a certain number of possible amino acid combinations that can come up to this range +/- 1ppm (0.0017 amu). If we look at the series of the fragmentation I highlighted above from left to right we see that the gap between the first two amino acids is exactly the mass of the amino acid Alanine (A). The gap between the next two is exactly that of Valine (V). The next one is trickier. Leucine and Isoleucine have the exact same mass, so we can't tell the difference between them in normal mass spectrometry. The following gap is straight-forward, as it is the exact mass of Glutamine (Q, the inventors of amino acid nomenclature ran out of letters...).

Therefore this peptide with an exact mass of 1755.947 that ends in lysine (L) has a sequence inside it that goes AV (I or L) Q. Fortunately we get the fragmentation from both ends. By looking at the overlap we can re-assemble this whole peptide.

This approach is known as *de novo* (in Latin, this roughly means "from the beginning," see, High School Latin did pay off. Thanks, Father Hogan!) sequencing. This approach is kind of a personal obsession of mine and I commonly evaluate new software that emerges to perform these analyses. I start with peptides that I know and I see if the *de novo* software can come up with the correct sequence. If it can, reliably, then I can trust the software to give me the correct answers for peptides and proteins that I don't know.

This information can be invaluable for the study of extinct species. Once we have the protein sequence for our organism of interest we can compare these sequences to those of current organisms by a process known as BLAST. BLAST stands for Basic Local Alignment Search Tool and is a free service provided by the National Institute of Health. You can enter any protein sequence and search it versus any known protein sequence on earth. In this way, you can find out that the bit of skin you found in that block of amber is closely related to the protein of alligators, frogs and chickens. You might be one step closer to that cage of velociraptors you've always dreamed of.

(Image courtesy of Matt Martyniuk)

References:

1.) Savitski, Mikhail M., et al. "Proteomics-grade de novo sequencing approach." Journal of proteome research 4.6 (2005): 2348-2354.
2.) Waridel, Patrice, et al. "Sequence similarity-driven proteomics in organisms with unknown genomes by LC-MS/MS and automated de novo sequencing." Proteomics 7.14 (2007): 2318-2329.

22 DATA INDEPENDENT EXPERIMENTS

Wait? What? How can you do an experiment that is independent of data? This is the response a physician gave an acquaintance of mine when he described this type of experiment. Fortunately, it is simply an unfortunate naming convention based on the fact that shotgun proteomics experiments described previously are also called Data Dependent experiments. In these experiments the mass spectrometer looks for ions that look like peptides according to parameters that you gave the instrument and then selectively fragments those ions.

In a Data Independent Acquisition (DIA), the mass spec doesn't slow down enough to tell anything about the ion that it is looking at. It simply fragments them all.

There are many names for this type of experiment due to the fact that mass spec manufacturers have licensed these with catchy marketing terms. The best promoted terms are MSe, AIF and SWATH (these are all trademarked terms; please refer to the disclaimer page).

DIA experiments represent a dramatic shift in the paradigm of mass spectrometry of proteins. In data dependent experiments (DDA) our experiments are set up in a way that we can discover new proteins and protein modifications. In DIA we allow someone else to do this discovery work for us and then we use those results.

DIA works like this:

1) Peptides are separated by HPLC

2) The mass spectrometer scans all of the peptides in a mass range. Sometimes this mass range is large, and other times it is much smaller. Common widths are about 25 amu

3) The mass spectrometer then fragments all of the ions within that mass range and collects data on all of the fragments

4) The data from the time the peptides came out of the HPLC, the unfragmented masses and the fragment masses are all compared against libraries of known peptides.

As mentioned in step 4, peptide spectral libraries are an essential component of DIA. Someone somewhere had to discover a peptide and accurately identify it. Then they took the time to save it to a library database with all the relevant information -- including the time the peptide takes to come off of HPLC separation, the peptide exact mass and the peptide fragment masses. If you are looking at an organism that no one ever studied before, you can't do DIA. In fact, if your organism hasn't been studied thoroughly by people who knew what they were doing, you can't do DIA either.

There are certainly weaknesses to this approach. What if the library you are searching is wrong? Then all of your analysis will be wrong. What if they used a different fragmentation type? Then your data will not match theirs and your analysis will be wrong.

Fortunately good spectral libraries are being created all the time. In February of 2013, a massive global team of researchers reported the release of the complete spectral libraries for baker's yeast. This freely downloadable database includes all of the information I noted above for every peptide from every protein that yeast can produce. The complete library can be obtained from a resource called the Peptide Atlas.

Researchers at the National Institute of Standards and Technology have developed huge spectral libraries for basically every compound they have been able to purify so far on earth. Everything from drugs to minerals to materials from meteors have been studied and they are working on assembling libraries for proteins and peptides. The construction of disease specific protein libraries is also a big emphasis for the Biomarkers Research Initiatives in Mass Spectrometry (BRIMS) center in Cambridge, MA.

The real strength of the DIA approach is that by collecting fragmentation data on every ion we can search the libraries of today and if 5 years from now we have much better and more thorough libraries, we can go back to this same data set and search it with the new libraries, with no need to re-run the sample on the mass spectrometer.

Reference:

Picotti, P. et al. A complete mass-spectrometric map of the yeast proteome applied to quantitative trait analysis. Nature advance online publication (20 January 2013)

23 HOW LOW CAN WE GO?

I've mentioned before that mass specs are sensitive devices. Arguably, we are talking about the most sensitive devices ever constructed. We are measuring the masses of individual atoms with the selectivity to easily distinguish minute nuclear differences. But we're often talking about thousands if not millions of identical molecules that we've enriched or selected in some way.

A good question for mass spectrometry is what are the limits of sensitivity? Fortunately, this is a competitive battle ground. As a mass spec manufacturer, there is a whole lot of money to be made if you have the most sensitive instrument on the market. Due to this the sensitivity is increasing all the time. It would be useless for me to write a number with a whole lot of zeroes behind the decimal place write now. By the time this has made it through the proofreading process, the editors, and the printing (or, more likely, the Kindle-ization process) to get to your hands, there would be more zeroes in the instruments that are for sale right now.

To get things into scale, though, I'm going to refer to a recent study that illustrates the level of sensitivity of today's approaches. This study involved the analysis of molecules from one single cell of an organism. That is the level of sensitivity that we are talking about.

In this 2013 study reported in the Proceedings of the National Academy of Sciences, an international team of scientists report a method to separate yeast cells into single cells on a plate. They were then able to use laser ionization and time-of-flight mass spectrometry to accurately identify and quantify the levels of sugars in an individual yeast cell. These researchers described the sensitivity they were able to achieve as 10 femto-mols, or 0.000 000 000 000 010 moles. Again, as I mentioned earlier, it takes a while to publish things. This study was published nearly 1.5 years ago while I am writing this, but by even the time of this writing I know that femto-mols is not a unit that manufacturers are bragging about these days.

Single cell analysis isn't going away either. The RIKEN institute in Saitama, Japan recently opened a center devoted purely to the analysis of single cells with mass spec technologies.

References:

1) Alfredo J. Ibáñez, Stephan R. Fagerer, Anna Mareike Schmidt, Pawel L. Urban, Konstantins Jefimovs, Philipp Geiger, Reinhard Dechant, Matthias Heinemann, and Renato Zenobi; Mass spectrometry-based metabolomics of single yeast cells, PNAS 2013 110 (22) 8790-8794

24 HOW HIGH CAN WE GO?

Building on the last chapter, I think the next question to consider is how high can we go? In other words, what is the biggest molecules that we can study by mass spectrometry? This is interesting because big molecules are exactly the opposite of what mass spectrometry started out as. In the beginning it was used purely to study the properties of groups of atoms. With John Fenn's discovery of electrospray ionization he was one of the first to successfully obtain mass spectra of biological molecules such as peptides and proteins.

In the chapter on ADCs I talked about studying antibodies, which are pretty big proteins. For most instruments at the point of this writing, this represents the real upper limit for mass spectrometry. And even today the analysis of antibodies (150,000 amu) is not an easy experiment. It still lies in the realm of experts. This fact helps keep a roof over my head as helping scientists with antibody analysis is a large part of my day job.

The Heck Lab in Utrecht, Netherlands has been pushing the boundaries of large molecule analysis. As I mentioned in the chapter on immunoprecipitation, sometimes we want to know what proteins are interacting together. In the scheme that I described, we pull down the proteins we are interested in with an antibody, digest them and then analyze the digested peptides to discover the proteins present. The Heck lab wants to leave out the digestion step.

In 2013, this team reported the use of a modified Orbitrap analyzer that allowed them to do just that. By changing the vacuum and by making other internal modifications to the instrument the team demonstrated that they could study protein complexes made up of proteins of exceeding 1,000,000 (1 million) atomic mass units. Prior to this study, no one had ever even approached the 1 million amu mark. Fortunately for everyone interested in studies of this type, the modified Orbitrap has been developed and was released as a complete product by Thermo Scientific around November of 2013.

And this is fortunate for science, in general, as looking at proteins in their undigested form has revealed how little we know about proteins in general. Ovalbumin is the main protein that makes up egg white. It has been used as a standard in nearly every protein lab since the beginning of time. Despite its inclusion in hundreds of research reports and papers, last year the Heck lab showed that we know almost nothing about it. Using a technique called native mass spectrometry where great care is taken to keep the protein in its natural state, the Heck lab demonstrated that ovalbumin isn't a pure protein at all. What we thought was a single protein is actually a mix of at least 59 different isoforms that have been simply destroyed or are indistinguishable by every other biochemical technique.

With the release of the Heck Orbitrap (called the Exactive EMR) as a commercial product, I fully expect a wave of new data that will open our eyes to the real workings of proteins and other large molecules.

References:

1) Nature Methods 9, 1084–1086 (2012) doi:10.1038/nmeth.2208;Received 20 June 2012 Accepted 17 September 2012

2) Anal. Chem., 2013, 85 (24), pp 12037–12045; DOI: 10.1021/ ac403057

25 MIND READING

Biochemistry is the field of science that looks at the building blocks of life. It deals with how biological molecules work together to make a plant or a person or even the majestic pug. Every movement we make is a complex biological process involving neural discharges that are translated into energy exchange with motor proteins and involves millions of interacting molecules.

In what might be considered a theme from a bad Science Fiction movie, there is mounting evidence that we can use mass spectrometry to tell what is happening in the brain in response to stimuli.

Two studies stand out to me in particular. The first involves using proteomics to study the effect of fear in a mouse brain. In this study, first published in January of 2014 a team in Colorado used a common psychological tool called fear conditioning on one set of mice. A control set was not trained in this way.

I'm no expert on psychological techniques. I understand that fear conditioning is similar to Pavlovian conditioning except that the mouse was taught to attribute a certain stimuli with an electrical shock. In this manner you can teach a mouse to fear that stimulus.

What happens when we pull out the mouse brain and compare it to the brain of a mouse that we haven't spent all day scaring and zapping? We can find measurable differences in the phospho-proteome of the two brains.

Phosphorylation is the act of one protein to add a phosphate group (PO_4) to another protein. This process can be almost instantaneous and is one of the ways that cells can quickly respond to stress and stimuli. Adding a phosphate to a protein can turn that protein off or turn it on or even change the other proteins that it interacts with. Analysis of what proteins are phosphorylated in a cell and when is called phosphoproteomics.

Phosphoproteomics is tricky. Phosphorylation is so instantaneous in a cell because phosphos are relatively easy for control proteins to put onto and pull off of proteins. Unfortunately the also fall off of proteins and peptides during mass spec analysis. Great care must be taken even during the sample preparation techniques in order to not disturb the actual phosphoproteome. So-called "soft fragmentation" methods are often necessary to accurately capture the location of the phosphorylation on proteins and peptides.

If you can get through these challenges, however, phosphoproteomics gives us insight into what is happening in the cell during that instant. We have only scratched the surface of the full understanding of what phosphorylations mean in even the yeast cell, let alone what is happening in complex organisms like mice and humans.

What did this mouse study discover? It found that we could clearly see the phosphoproteomic footprint of chemical reactions occurring in the mouse brain that could be attributed to fear. Stop for a second and consider the applications of this. Could we start with a deceased mouse and then determine by looking for these phosphorylated proteins to tell if the mouse was afraid when it died? Could we tell how afraid? Extending this, could we do the same thing for a victim found at a crime scene? Could this be a standard step in a future coroner's analysis?

What other things could we tell if we knew what phosphoproteins to look for? Sure, I'm reaching, but we have to consider these possibilities.

A second recent study looked at mouse brains when mice were treated with hallucinogenic drugs. Looking at the impact of these compounds on the phosphorylation of proteins gives us another glimpse into brain chemistry in mice. Unfortunately, I'm sure there are a limited number of applications of the mouse brain that directly translate into human beings. Beyond scaring them and giving them LSD, I can't think of anything else. I can't exactly check to see if the phosphoproteome of mouse brains change when mice are lying and this procedure is a bit too invasive for human studies as it appears to involve the removal of the entire brain.

Phosphoproteomics is another rapidly expanding field of mass spectrometry. Thorough studies of this type on different organisms and during the stages of different diseases are commonplace in the best scientific journals these days.

References

1) Mol Cell Proteomics. 2014 Apr;13(4):919-37. doi: 10.1074/mcp.M113.035568. Epub 2014 Jan 27.
2) Mol Cell Proteomics. 2014 May;13(5):1273-85. doi: 10.1074/mcp.M113.036558

26 VACCINE RESEARCH

These days I don't do a lot of my own research. A couple years ago I chose to take a job where I help other people do their research. In order to honestly keep the word "scientist" on my business cards, I force myself to continue to do as much science as I can. In the little bit of time I have to focus on actual research, I spend assisting a malaria proteomics lab in Rockville, MD.

It is truly hard to get a good number on the impact of malaria in the world, but the World Health Organization estimates that between 600,000 and 1.2 million people in the world died from malaria in the world in 2010. It is hard to get numbers because most of these deaths were in the poorest African countries. Most of these deaths were children who didn't live long enough to be recorded by the informal censuses the outside world has for these places.

The organism responsible for malaria is a complex protozoan called *Plasmodium falciparum* (Pf). Pf has a complex life cycle in which it goes through various developmental cycles and is transmitted by mosquitoes which move the parasite from infected to uninfected individuals.

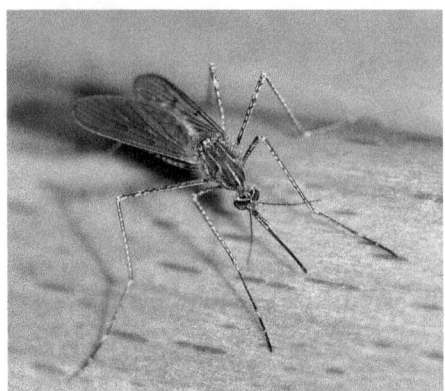

As you might imagine, malaria is probably one of the last things you would want contract while pregnant. Surprisingly, however, sometimes that is a good thing.

For decades, Michal Fried and Patrick Duffy have studied malaria in central Africa, with a particular focus on women who contract malaria while pregnant. Over the course of their research they have discovered this trend:

Women who are pregnant and contract malaria (let's call them P1) develop severe malaria symptoms and almost 100% of the time lose the child they are carrying.

P1 women who survive and get pregnant again (let's call this group P2) are almost completely immune to malaria, and their babies nearly always survive.

What they have discovered is that the placentas of P2 women possess a special property -- the malaria parasite sticks to the placenta and can't cause harm to the woman or the baby. To determine why this happens, Michal and Patrick have collected countless samples of blood and even placental samples from both P1 and P2 women.

Creating this sample collection has been a painstaking process involving trudging across villages in some of the most remote countries in the world. This work has put them in some precarious situations. A few years ago they were present when the military of Mali overthrew the government. Both they, and a number of their students and staff were detained in the country while the new government decided what to do with them. Their work has been noted and they have been the recipients of both government funding and a generous support grant from the Bill and Melinda Gates Foundation.

Michal's focus in her lab, the Molecular Pathogenesis and Biomarkers facility at the NIH, is on the proteomics of these P1 and P2 samples. A long line of mass spectrometrists (of which I was once one) have used technique after technique to determine why the parasite sticks to the P2 placentas. As the technology in mass spectrometry has improved, Michal has adopted any and every technique possible to find this protein that might lead to a cheap and effective malaria vaccine.

Currently, my good friend Dr. Patricia Gonzales performs the mass spec analysis of these samples using at cutting edge Orbitrap instrument. I am confident that it is only a matter of time before these samples finally reveal their secrets to this lab. I only hope it is sooner, rather than later.

References:

1.) Fried, Michal, et al. "Maternal antibodies block malaria." *Nature* 395.6705 (1998): 851-852.
2.) Duffy, P. E., and M. Fried. "Malaria in the pregnant woman." Malaria: Drugs, Disease and Post-Genomic Biology. Springer Berlin Heidelberg, 2005. 169-200.

DISCLAIMER STATEMENTS

The products and technologies mentioned in this text are purely the properties of the manufacturers of these products

Screenshots from instruments and software in this text are from Thermo Fisher Scientific. All applicable trademarks and copyrights of these software and instruments are the sole property of Thermo Fisher Scientific. These screenshots were taken to illustrate scientific techniques.

Orbitrap and Exactive are registered trademarks of Thermo Fisher Scientific

SWATH is a registered trademark of SCIEX and Applied Biosystems

MSe is a registered trademarked property of the Waters Corporation

Neither I, nor my publisher, own any rights to television shows or fictional characters mentioned in this text. These are pop culture. Chill out.

Images of and taken by the Curiosity Mars Rover are property of the U.S. government and legally distributable by the author as a U.S. citizen. Reproduction of these images by citizens of other nations is prohibited.

All effort has been made to rightfully attribute the research, illustrations and techniques to their appropriate inventors and right holders.

Please do not sue this author.

ABOUT THE AUTHOR

Ben Orsburn received his Ph.D. in Biological Sciences from Virginia Tech in 2009. Due to a history of rampant overconfidence and inexhaustible enthusiasm he has tried just about every mass spectrometry experiment out there at least once, with varying levels of success. He is the author of the surprisingly popular blog: News in Proteomics Research, where he rambles about methods, advances in the field and pugs to a large and very supportive audience. He currently lives in Baltimore, MD and wrote a lot of this in the company of two old snoring pug dogs.

www.ingramcontent.com/pod-product-compliance
Lightning Source LLC
Chambersburg PA
CBHW051729170526
45167CB00002B/861